Lecture Notes in Computer Science

Lecture Notes in Artificial Intelligence   13824

Founding Editor

Jörg Siekmann

Series Editors

Randy Goebel, *University of Alberta, Edmonton, Canada*
Wolfgang Wahlster, *DFKI, Berlin, Germany*
Zhi-Hua Zhou, *Nanjing University, Nanjing, China*

The series Lecture Notes in Artificial Intelligence (LNAI) was established in 1988 as a topical subseries of LNCS devoted to artificial intelligence.

The series publishes state-of-the-art research results at a high level. As with the LNCS mother series, the mission of the series is to serve the international R & D community by providing an invaluable service, mainly focused on the publication of conference and workshop proceedings and postproceedings.

Makoto Yokoo · Hong Qiao ·
Yevgeniy Vorobeychik · Jianye Hao (Eds.)

Distributed
Artificial Intelligence

4th International Conference, DAI 2022
Tianjin, China, December 15–17, 2022
Proceedings

Springer

Editors
Makoto Yokoo
Kyushu University
Kyushu, Japan

Hong Qiao
University of Science and Technology
Beijing, China

Yevgeniy Vorobeychik
Washington University in St. Louis
Louis, MO, USA

Jianye Hao
Tianjin University
Tianjin, China

ISSN 0302-9743   ISSN 1611-3349 (electronic)
Lecture Notes in Artificial Intelligence
ISBN 978-3-031-25548-9   ISBN 978-3-031-25549-6 (eBook)
https://doi.org/10.1007/978-3-031-25549-6

LNCS Sublibrary: SL7 – Artificial Intelligence

This Springer imprint is published by the registered company Springer Nature Switzerland AG
The registered company address is: Gewerbestrasse 11, 6330 Cham, Switzerland

Preface

Over past decade, there has been tremendous development in the artificial intelligence (AI) community. Therefore, a new conference, the International Conference on Distributed Artificial Intelligence (DAI), has been organized since 2019. DAI aims at bringing together international researchers and practitioners in many related areas. This year, we received 16 submissions in total. Each paper was assigned to four Program Committee (PC) members, and each received at least two reviews, and on average three reviews. With the reviews, the decisions were made based on the discussion and consensus of the Program Committee with the PC chairs. The topics of the accepted papers included reinforcement learning, multi-agent learning, distributed learning systems, deep learning, and applications of game theory. We were also delighted to have as invited speakers Peter Stone (University of Texas), Katja Hofmann (Microsoft Research Cambridge), Wotao Yin (University of California), Zhiwei (Tony) Qin (Didi Research America), Dacheng Tao (Jingdong Institute of Discovery). Finally, we would like to sincerely thank the conference committee and the Program Committee for their great work.

December 2022

Makoto Yokoo
Hong Qiao
Yevgeniy Vorobeychik
Jianye Hao

Organization

General Chairs

Makoto Yokoo   Kyushu University, Japan
Hong Qiao   University of Science and Technology Beijing,
               China

Program Committee Chairs

Yevgeniy Vorobeychik   Washington University in St. Louis, USA
Jianye Hao   Tianjin University, China

Program Committee

Priel Levy	Bar-Ilan University, Israel
Reuth Mirsky	University of Texas, Austin, USA
Ofra Amir	Technion, Israel
Meenal Chhabra	Square, USA
Mithun Chakraborty	University of Michigan, USA
Allen Lavoie	Google, USA
Zhuoshu Li	Google, USA
Sebastian Stein	University of Southampton, UK
Valentin Robu	Centrum Wiskunde & Informatica, The Netherlands
Haifeng Xu	University of Virginia, USA
Haris Aziz	UNSW, Australia
Weiran Shen	Renmin University, China
Frans Oliehoek	TU Delft, The Netherlands
Jilles Dibangoye	INSA Lyon, France
Shimon Whiteson	University of Oxford, UK
Amanda Prorok	University of Cambridge, UK
Akshat Kumar	Singapore Management University, Singapore
Roi Yehoshua	Northeastern University, USA
Maria Gini	University of Minnesota, USA
Jen Jen Chung	ETH Zürich, Switzerland
Zheng Tian	UCL, UK

Yaodong Yang	KCL, UK
Haitham Bou-Ammar	Huawei, UK
Stefano V. Albrecht	University of Edinburgh, UK
Matthew E. Taylor	University of Alberta, Canada
Ming Zhou	SJTU, China
Zongzhang Zhang	Nanjing University, China
Guifei Jiang	Nankai University
Chongjun Wang	Nanjing University, China
Jianye Hao	Tianjing University, China
Chongjie Zhang	Tsinghua University, China
Feng Wu	University of Science and Technology of China, China
Chao Yu	Sun Yat-sen University, China
Bo An	Nanyang Technological University, China
Noam Hazon	Ariel University, Israel
Amos Azaria	Ariel University, Israel
Jacopo Banfi	MIT, USA
Alberto Quattrini Li	Dartmouth University, USA
Joydeep Biswas	UT Austin, USA
Roni Stern	Ben Gurion University, Israel
Roie Zivan	Ben Gurion University, Israel
Harel Yedidsion	UT Austin, USA
Long Tran-Thanh	University of Warwick, UK
Weinan Zhang	Shanghai Jiao Tong University, China
Siqi Chen	Tianjin University, China

Contents

A Distributed RBF-Assisted Differential Evolution for Distributed Expensive Constrained Optimization

Feng-Feng Wei, Xiao-Qi Guo, Wen-Jin Qiu, Tai-You Chen,
and Wei-Neng Chen$^{(\boxtimes)}$

South China University of Technology, Guangzhou, China
cschenwn@scut.edu.cn

Abstract. With the development of Internet of things and distributed computing techniques, distributed and expensive constrained optimization problems (DECOPs) have emerged in the industry. DECOPs refer to optimization problems with objective and constraint functions that are computationally expensive and can only be evaluated on multiple agents of distributed networks. In DECOPs, the raw data of each agent cannot be transmitted to other agents, but only objective or constraint value of a solution can be evaluated, resulting in the incomplete data on each agent. This paper proposes a distributed RBF-assisted differential evolution (DRADE) algorithm for solving DECOPs. In DRADE, we added a master agent to the distributed networks of DECOPs, connecting work agents that can evaluate objective or constraint values of candidate solutions to the master agent in a star topology. The proposed algorithm is composed of candidate generation and selection on master agent and radial basis function (RBF) management on work agents. In candidate generation and selection, differential evolution serves as an optimizer to generate candidate solutions assisted by RBF models received from work agents to replace expensive evaluations of candidate solutions in the master agent. In RBF management, each work agent constructs and updates a RBF model with its own data, which are updated by samples selected from candidate solutions received from the master agent and their expensively evaluated values. Statistical results and analysis of experiments carried out on benchmark test functions and engineering problems show that DRADE has superior performance than compared state-of-the-art SAEAs.

Keywords: Distributed optimization · Surrogate-assisted evolutionary algorithm · Expensive constrained optimization · Differential evolution

---

This work was supported in part by the National Natural Science Foundation of China under Grant 61976093. The research team was supported by the Guangdong Natural Science Foundation Research Team No. 2018B030312003 and State Key Laboratory of Subtropical Building Science.

M. Yokoo et al. (Eds.): DAI 2022, LNAI 13824, pp. 1–14, 2023.
https://doi.org/10.1007/978-3-031-25549-6_1

1 Introduction

Constrained optimization problems (COPs) are often encountered in industrial
applications in many real-world areas, such as medical [23], aeronautical design
[12], industrial manufacturing [5], resource [8] and so on. Evolutionary algorithms (EAs) are widely used for solving COPs [17,22] due to their powerful
search abilities. The constraint-handing techniques in EAs are generally divided
into four categories, namely penalty function methods [20], repairing methods
[4], methods based on feasibility rule [18] and methods based on multi-objective
optimization algorithms [10]. It is noteworthy that objectives and constraints of
many COPs in industry are computationally expensive and cannot be modeled
as explicit mathematical expressions. This kind of COPs is named as expensive
ECOPs (ECOPs). Existing EAs that employ the constraint handling techniques
mentioned above often need to iteratively evaluate both the objective and constraint values of candidate solutions, which takes a long time to solve ECOPs.

To address the above issues, quite a few efforts have been made to use surrogate models to replace part of expensive evaluations of objectives or/and constraints. This kind of methods are usually named surrogate-assisted evolutionary algorithms (SAEAs). For single-objective ECOPs, Su et al. [20] proposed
a hybrid surrogate-based-constrained optimization method, which replaces the
objective function with penalty function by a Kriging model in the first phase and
approximates both objective and constraints with radial basis function (RBF)
models in the second phase. Rahi et al. [18] devised a surrogate-assisted partial-
evaluation-based EA to address ECOPs. Handoko et al. [6] adopted a support
vector machine to construct a feasibility structure model to judge the feasibility
rule between two solutions in memetic algorithms in ECOPs. Wang et al. [24]
designed a global and local surrogate-assisted differential evolution for ECOPs
with inequality constraints, in which a generalized regression neural network is
served as global surrogate model and RBF is used as local surrogate model to
create new solutions. For multi-objective combinatorial ECOPs, Wang et al. [23]
proposed a random forest-assisted EA.

Almost all of the above SAEAs are designed for centralized ECOPs, but few of
them consider the situation that objectives and constraints are also distributed.
In fact, objectives and constraints of many real-world ECOPs need to be acquired
by different ways on multiple agents, which is named as distributed and expensive constrained optimization problems (DECOPs). DECOPs are common in the
industry but have not received much attention in academia. For example, aircraft
design subjects to material properties, energy supply requirements, aerodynam-
ics principles and so on. Material properties such as static or dynamic stiffness
of aircraft materials require finite element analysis (FEA) simulation [7]. The
motor drive for aircraft energy supply demands high reliability and availability,
which should be verified by FEA [1]. Airfoil shape parameters should meet aero-
dynamic principles [12]. These constraints are acquired by different simulation
tools, which may be evaluated on different agents. Since the raw data of each
agent cannot be transmitted to other agents, we can only obtain one objective
value or constraint value of a candidate solution on each agent, leading to a

new challenge in DECOPs with incomplete information. Incomplete information makes it difficult to distinguish among individuals in the population of each agent, thus it is hard to guide the population to evolve and generate promising candidates. Therefore, it is worth researching how to select suitable candidates in the case of distributed evaluations of objective and constraints in DECOPs, which is of great significance for improving algorithm performance.

Taking account of the above challenges, we propose a distributed RBF-assisted differential evolution (DRADE) for solving DECOPs. Contributions of the proposed algorithm are described as follows.

Firstly, we add a master agent in the distributed network of each single-objective DECOP to construct a star distributed topology. The master agent needs to coordinate all work agents to solve DECOPs, but has no ability to evaluate the objective or constraints. Work agents are responsible for expensive evaluations of the objective or one of constraints, and can communicate with the master agent to help solve problems.

Secondly, the proposed DRADE consists of candidate generation and selection on master agent and RBF management on work agents. The RBF model on each work agent is transmitted to the master agent to avoid transmission of raw data so that data privacy can be protected. Besides, it serves as surrogate model to assist evolutionary optimization on the master agent and save the number of real evaluations. The RBF model on each work agent serves as a surrogate model to replace the data transmission and is transmitted to the master agent to assist evolutionary optimization, which can not only save the number of expensive evaluations but also protect data privacy. Differential evolution (DE) acts as a search engine and uses approximated estimation of RBF models to generate candidate solutions, which are fed back to work agents for expensive evaluations and updating RBF models.

The remainder of this paper is organized as follows. Section 2 briefly introduces the definition of DECOPs. Section 3 elaborates the proposed DRADE for DECOPs in detail. Experimental studies and analyses on benchmark suites and ceramic formular optimization problems of the proposed algorithm are presented in Sect. 4. Finally, Sect. 5 concludes this paper.

2 Distributed and Expensive Constrained Optimization

In this paper we consider to solve single-objective DECOPs, in which evaluations of objectives and constraints are computationally expensive and need to be acquired in distributed ways. The two characteristics of expensive and distributed often occur in industrial problems at the same time. To be specific, many expensive evaluations or tests in the industry often need to be completed in specific institutions with specific equipment, so an optimization problem with multiple constraints becomes a distributed problem.

For example, the optimization of vehicle parameters [3,18] in automobile manufacturing needs to meet constraints such as safety performance, structural performance, and air resistance, and these three types of assessments need to be completed by corresponding professional institutions, respectively. Therefore, the process of optimizing the parameters needs to call multiple organizations to complete through communication.

Without loss of generality, a minimization single-objective DECOP can be formulated as follows:

$$\min_{\boldsymbol{x} \in R^D} \hat{f}^{a_0}(\boldsymbol{x})$$
$$s.t. \ \hat{C}_k^{a_k}(\boldsymbol{x}) \leq 0, \ k = 1, 2, ..., l \tag{1}$$
$$g_i(\boldsymbol{x}) \leq 0, \ i = 1, 2, ..., m$$
$$h_j(\boldsymbol{x}) = 0, \ j = 1, 2, ..., n$$

where $\boldsymbol{x}$ is the decision variable vector and each variable is bounded by $x_i \in [lb_{x^i}, ub_{x^i}]$, $i = 1, 2, ..., D$. There are three types of constraints, which are l expensive inequality constraints $\hat{C}_k^{a_k}(\boldsymbol{x})$, m inexpensive inequality constraints $g_i(\boldsymbol{x})$, and n inexpensive equality constraints $h_j(\boldsymbol{x})$. Because the evaluations of objective function and expensive constraints are computational expensive, they need to be computed in specific agents. The superscript of $\hat{f}^{a_0}$ or $\hat{C}_k^{a_k}$ represent the agent that computes it. In this definition, the data and evaluation information is private to other agents. Therefore, objective function $\hat{f}^{a_0}$ is computed in agent a_0 only, and constraint $\hat{C}_k^{a_k}$ is evaluated in agent a_k only.

DECOPs have posed two challenges to optimization methods. On the one hand, it is not practical to evaluate solutions frequently in the algorithm, considering the economic cost and the time cost. On the other hand, due to the distribution of problem information and data privacy, the optimization requires multiple agents to complete it cooperatively.

3   Distributed RBF-Assisted Differential Evolution

This section elaborates the proposed DRADE for DECOPs in detail. Specifically, the framework of DRADE is firstly introduced. Then, two important parts, RBF management and candidate generation and selection are presented to facilitate DRADE.

3.1   DRADE Framework

DRADE is a distributed algorithm. It adopts a master-slave model on a star topology. The framework of DRADE is shown in Fig. 1. There are 1 master and $NC+1$ work agents, in which NC is the number of distributed and expensive constraints. Work agent #1, #2, ..., #NC is able to evaluate the $1^{th}, 2^{nd}, ..., NC^{th}$ constraint respectively and work agent #$NC + 1$ has the ability to evaluate the objective values of candidate solutions. The master takes charge of evolution to search for global optima through DE. Each work agent has the ability of expensive evaluation for the objective or one constraint. It can also train an RBF model using its private historical data. The master has no ability of any evaluations. It receives RBF models from work agents to predict the objective and

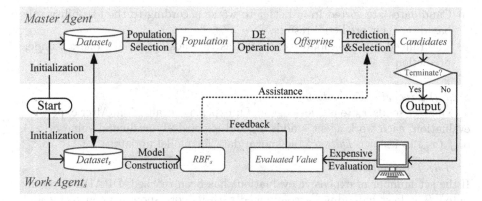

Fig. 1. Framework of DRADE

constraints values of generated candidates. Only a few promising candidates are sent to work agents for expensive evaluations. Work agents are mutually confidential and they only can communicate with the master. The main procedures are elaborated as follows.

Step 1) Initialization.

a) The master initializes an empty $Dataset_0$ to save expensively evaluated candidates. Besides, a population P with only decision variables x_i for evolution, where $x_i = \{x_i^1, x_i^2, ..., x_i^D\}$ is the decision variable in D dimensional search space. $i = 1, 2, ..., NP$ is the index of individuals and NP is the population size.

b) The master asks all work agents to expensively evaluate constraints $G(x_i) = \{g_1(x_i), g_2(x_i)..., g_{NC}(x_i\}$ and objective $F(x_i)$ of the population P. They are added to the $Dataset_0$ as known data.

c) Each work agent s initializes a $Dataset_s$ with NS historical data, which include decision variables $X_s = \{x_1, x_2, ..., x_{NS}\}$ and corresponding evaluated constraints $Y_s = g_s(X_s)$ or objective $Y_s = f_s(X_s)$.

Step 2) RBF Management.
Each work agent s uses all data $\{X_s, Y_s\}$ in its own dataset to train or update its RBF regression model $\mathcal{M}_s$. Once $\mathcal{M}_s$ is updated, work agent s sends it to the master.

Step 3) Candidates Generation.
    Based on the feasibility rule which is introduced in Subsect. 3.3, the best NP individuals are selection from A to form the population P. Three composite DE evolution operators are conducted on P to generate $NP \times 3$ candidates $x'_{i,j}, i = 1, 2, ..., NP, j = 1, 2, 3.$

Step 4) Candidates Preselection.

a) Based on received RBF models $\mathcal{M}_s, s = 1, 2, ..., NC + 1$, the master predicts constraints and the objective of generated candidates $\{\hat{G}(x'_i), \hat{F}(x'_i)\}$, where $\hat{G}(x'_i) = \hat{g}_1(x'_1), \hat{g}_2(x'_2), ..., \hat{g}_{NC}(x'_{NC})$

b) Candidates are sorted from better to worse according to the feasibility rule based on their predicted results.

c) The ranked first candidate is selected as the most promising one x_p under prediction.

Step 5) Expensive Evaluation.
The master sends x_p to all work agents for expensive evaluation. After expensive evaluation, each work agent s adds x_p and expensively evaluated value $g_s(x_p)$ or $f_s(x_p)$ to its own $Dataset_s$ as known data for model update.

Step 6) Termination.
If the set number of expensive evaluations has been reached, DRADE is terminated. The best individual, accompanied with its corresponding constraint and objective values are output.

Otherwise, DRADE goes to *Step 2)* to continue evolution.

3.2   RBF Management

Radial basis function (RBF) uses a series of basis functions for interpolation. Due to its high efficiency and flexibility, RBF has become a widely used model in SAEAs [9].

At the start of the algorithm, each work agent s uses data $\{X_s, Y_s\}$ in its $Dataset_s$ to build an RBF regression model $\mathcal{M}_s$ and sends it to the master. Specifically, the approximation scheme is shown as follows:

$$\hat{G}(x) or \hat{F}(x) = \sum_{j=1}^{n} \lambda_j \phi(||x - x_j||) \qquad (2)$$

in which $x_j \in X_s$ are training data in $Dataset_s$. n is the total number of training data, which is initialized as NS and increased during the evolution. x is the individual to be predict. λ_j is the weight parameter to be trained. $|| \cdot ||$ is the Euclidean distance between x and x_j. ϕ is the radial basis function. It adopts Gaussian function $\exp(-\frac{x-x_j}{2\delta})$, where δ is the smoothing parameter.

During the evolution, work agents update their $\mathcal{M}$ once new individuals are expensively evaluated. In other words, models are updated in each generation to improve approximation accuracy and sent to the master to assist evolution.

3.3   Candidate Generation and Selection

Differential evolution (DE) is an effective EC algorithm with powerful global search ability and has been widely applied to solve complex optimization problems [9]. After initialization and receiving the trained models $\mathcal{M} = \{\mathcal{M}_1, \mathcal{M}_2, ..., \mathcal{M}_{NC+1}\}$ from work agents, the master conducts composite DE evolution on the population. For each individual x_i in the population P, the following three composite DE mutation operators are applied to generate candidates.

DE/current-to-rand/1

$$v_{i,1}^{t+1} = x_i^t + r \times (x_{r1}^t - x_i^t) + F \times (x_{r2}^t - x_{r3}^t) \tag{3}$$

DE/current-to-best/1

$$v_{i,2}^{t+1} = x_i^t + F \times (x_{cbest}^t - x_{r1}^t) + F \times (x_{r2}^t - x_{r3}^t) \tag{4}$$

DE/best/1
$$v_{i,3}^{t+1} = x_{cbest}^t + F \times (x_{r1}^t - x_{r2}^t) \tag{5}$$

where t is the evolution generation. $v_{i,j}^{t+1}$ is the jth mutated trial of the ith individual in the $t+1$ generation and it is obvious that $j = 1, 2, 3$. r is a D dimensional random decimal vector. r_1, r_2, r_3 are three randomly selected individuals from the population, which are different from other and do not equal to i. $cbest$ is the index of the best individual in the current population. F is the mutation factor between $[0, 1]$ and set ahead of time.

After that, the mutated trials are conducted binomial crossover as follows:

$$v_{i,j}^{t+1,d} = \begin{cases} v_{i,j}^{t+1,d}, & \text{if } r < CR \text{ or } d = d_{rand} \\ x_i^{t,d}, & \text{otherwise} \end{cases} \qquad d = 1, 2, ..., D \tag{6}$$

where r is a random decimal between $[0, 1]$. CR is the crossover factor and set ahead of time. d_{rand} is a randomly selected dimension to make sure at least one dimension conducts crossover.

Finally, candidates $x_i', i = 1, 2, ..., NP \times 3$ are generated through boundary check, in which

$$x_{i,j}' = max(v_{i,j}^{t+1}, \textbf{lb}) \oplus min(v_{i,j}^{t+1}, \textbf{ub}) \tag{7}$$

In a word, if one dimension is larger or smaller than corresponding upper bound or lower bound, it is directly assigned as the corresponding bound value.

At last, candidate selection is conducted based on feasibility rule, which is widely used to compare quality of solutions. For any two solutions x_{r1} and x_{r2}, x_{r1} is judged as the better one if one of the following conditions is satisfied:

$$\begin{cases} f(x_{r1}) < f(x_{r2}) \wedge DV(x_{r1}) < 0 \wedge DV(x_{r2}) < 0 \\ DV(x_{r1}) < 0 \wedge DV(x_{r2}) > 0 \\ DV(x_{r1}) < DV(x_{r2}) \wedge DV(x_{r1}) > 0 \wedge DV(x_{r2}) > 0 \end{cases} \tag{8}$$

Here, DV represents the degree of constraint violation, which is defined as $DV(x) = \sum_{k=1}^{l} \hat{C}_k(x) + \sum_{i=1}^{m} g_i(x) + \sum_{j=1}^{n} |h_i(x)|$. If none of the three conditions is satisfied, x_{r2} is judged to be better than x_{r1}.

Based on predicted results, all candidates are sorted in descending order according to the feasibility rule. The ranked first candidate is selected as the most promising one under prediction, which is sent to all work agents for expensively evaluation.

4   Experiments and Analyses

This section conducts extensive experiments to demonstrate the satisfactory performance of DRADE. Firstly, test problems in two benchmark suites are introduced. Secondly, parameter and experiment settings are given. Thirdly, DRADE is compared with four centralized SAEAs to illustrate its competitive convergence and global search ability. Fourthly, an distributed and expensive optimization problem in industrial, the optimization of ceramic formula is tested to show the promising applications of DRADE in engineering optimization.

4.1   Test Suite

Two most widely used test suites for constrained optimization, CEC2006 [13] and CEC2010 [16] are adopted for extensive experiments. Out of reason that the feasible areas of equality constraints are extremely small, which is impossible for models to approximate and SAEAs for expensive constrained optimization generally do not handle them [6,19,24], we test problems with only inequality constraints. Problems and their characteristics in CEC2006 and CEC2010 are shown in Table 1 and Table 2 respectively.

Problems in CEC2006 have known optima whereas problems in CEC2010 are much more complicated and there is no optima information. They have various types of objective, which reflect different kinds of engineering problems. Feasible area ratio ρ, which is calculated through feasible areas $|F|$ divided by the whole search space $|S|$, of each problem is also given. I is the number of inequality constraints. Though constraints and objective of test suites are not expensive, they are frequently used for SAEA experiments [6,19,24].

Table 1. Test function of CEC2006

Name	D	The optimal	Type	$\rho = |F|/|S|$	I
g01	13	-15	quadratic	0.0111%	9
g02	20	-0.803619	nonlinear	99.9971%	2
g04	5	-30665.539	quadratic	52.1230%	6
g06	2	-6961.813876	cubic	0.0066%	2
g07	10	24.30620907	quadratic	0.0003%	8
g08	2	-0.095825042	nonlinear	0.8560%	2
g09	7	680.6300574	polynomial	0.5121%	4
g10	8	7049.248021	linear	0.0010%	6
g12	3	-1	quadratic	4.7713%	1
g16	5	-1.905155259	nonlinear	0.0204%	38
g19	15	32.6555929502	nonlinear	33.4761%	5
g24	2	-5.5080132716	linear	79.6556%	2

Table 2. Test function of CEC2010

Name	D	Type of objective	$\rho = |F|/|S|$	I
c01	10	Non-Separable	99.7689%	2
c07		Non-Separable	50.5123%	1
c08		Non-Separable	37.9512%	1
c13		Separable	0.0000%	3
c14		Non-Separable	0.3112%	3
c15		Non-Separable	0.3210%	3
c'01	30	Non-Separable	100.0000%	2
c'07		Non-Separable	50.3725%	1
c'08		Non-Separable	37.5278%	1
c'13		Separable	0.0000%	3
c'14		Non-Separable	0.6123%	3
c'15		Non-Separable	0.6023%	3

4.2   Parameter Setting

Population size $NP = 50$. Mutation factor F is selected from $\{0.6, 0.8, 1.0\}$ with the same probability $p_F = \frac{1}{3}$. Crossover factor CR is selected from $\{0.1, 0.2, 1.0\}$ with the same probability $p_{CR} = \frac{1}{3}$. The initialized dataset size of work agents $NS = 300$.

Due to the expensive cost of evaluations, the maximal number of expensive evaluations $FES = 1,000$ as most SAEAs set [15,21,25]. It should be noticed that in DECOPs, one evaluation of each work agent consumes one expensive evaluation budget. That means, full evaluation of a candidate consume $NC + 1$ evaluations, including evaluations of NC constraints and 1 objective.

Experiments are conducted on machine with 36 Intel® Xeon® CPU E5-2696 v3 @ 2.30 GHz processors. All results are averaged over 25 independent runs to avoid contingency.

4.3   Comparison with State-of-the-art SAEAs

It should be noticed that DECOP is a kind of emerging problem with the development of IoTs and distributed computing techniques. There are few studies focus on DECOPs. Therefore, we compare DRADE with four centralized SAEAs for expensive constrained optimization to illustrate its global optimization ability. The first one is Gaussian process surrogate-model-assisted evolutionary algorithm for computationally expensive inequality constrained optimization problems (GPEEC) [14]. The second one is Kriging-assisted teaching-learning-based optimization (KTLBO) [2]. The third one is multiple penalties and multiple local surrogates (MPMLS) [11] and the fourth one is surrogate-assisted classification-collaboration differential evolution (SACCDE) [26]. To make a fair comparison, all parameter and experiment settings are the same as their proposed papers.

Comparison results are shown in Table 3 and Table 4, in which *mean* is the average value of found-best feasible fitness over 25 independent runs and *std* is the corresponding standard variation. *rf* is the feasible ratio in the final population, averaged over 25 runs. It is calculated by the number of feasible individuals divided by population size. *rs* is the successful ratio which can successfully find feasible individuals over 25 runs. *p* is the Wilcoxon rank-sum statistical test value. $(+), (-), (\approx)$ means the compared algorithm is significantly better, worse or has no significant difference than DRADE at the significant level $\alpha = 0.05$. $(/)$ means there is no legal statistical test results out of reason that the compared algorithm cannot find any feasible solution. However, DRADE can find feasible solutions, which means DRADE is better than the compared algorithm. The following conclusions can be got from tables.

Table 3. Comparison result of DRADE with four SAEAs for CEC2006

		g01	g02	g04	g06	g07	g08	g09	g10	g12	g16	g19	g24	
GPE EC	mean	−6.3987	−0.2540	−30579	−6743.8	163.414	−0.0958	1681.31	11184.4	−1	NaN	1302.32	−5.5080	
		(−)	(−)	(−)	(−)	(−)	(≈)	(−)	(≈)	(≈)	(/)	(−)	(≈)	7 −
	std	0.68661	0.03617	24.2499	471.031	48.6336	0.00000	619.344	773.494	0	NaN	398.298	0	0+
	rf	0.806	1	1	1	0.84	1	0.989	1	1	0	1	1	4≈
	rs	1	1	1	1	1	1	1	1	1	0	1	1	1/
	p	1.42E−9	1.23E−6	1.42E−9	1.42E−9	3.21E−6	1.51E−1	1.31E−8	5.81E−2	2.07E−1	NaN	3.26E−9	9.12E−2	
KTL BO	mean	−0.7099	−0.3343	−29824	−4055.2	216.913	−0.0707	935.134	11502.2	−1	−1.2466	729.648	−5.1573	
		(−)	(≈)	(−)	(−)	(−)	(−)	(−)	(≈)	(≈)	(≈)	(−)	(−)	8 −
	std	1.29385	0.06603	262.969	1873.54	0	0.02523	96.1133	1925.96	0	0	293.488	0.20798	0+
	rf	0.81333	1	1	0.8	0.04	1	1	0.08	1	0.01333	1	1	4≈
	rs	0.88	1	1	0.8	0.04	1	1	0.08	1	0.04	1	1	0/
	p	4.85E−9	2.77E−1	1.42E−9	1.21E−8	1.43E−2	1.8E−10	1.07E−2	8.70E−1	2.07E−1	4.44E−2	1.13E−4	1.07E−9	
MPM LS	mean	−5.0994	−0.3232	−30597	−6954.5	1128.67	−0.0957	808.485	12588.9	−1	−1.5379	587.584	−5.507	
		(−)	(−)	(−)	(−)	(−)	(≈)	(+)	(−)	(≈)	(≈)	(−)	(−)	8 −
	std	0.98472	0.0562	37.1522	18.498	724.48	0	104.859	1580.48	0	0	146.176	0	1+
	rf	0.09866	0.912	0.85066	0.48	0.01266	0.80666	0.81533	0.23133	0.68933	0.002	0.874	0.62933	3≈
	rs	0.84	1	1	1	0.36	1	1	0.96	1	0.04	1	1	0/
	p	4.85E−9	1.16E−2	1.42E−9	1.42E−9	1.76E−5	7.59E−1	3.40E−3	3.95E−2	7.36E−2	5.73E−1	3.02E−5	4.0E−3	
SAC CDE	mean	−12.358	−0.3152	−30662	−6958.6	816.071	−0.0878	1279.25	15314.3	−0.999	−1.4489	357.717	−5.508	
		(−)	(−)	(+)	(−)	(−)	(−)	(−)	(−)	(≈)	(≈)	(−)	(≈)	8 −
	std	0.75282	0.06076	1.22942	10.0238	598.294	0.02211	1179.54	2732.55	0.00304	0.1216	141.115	0	1+
	rf	1	1	1	1	0.60533	1	1	0.22133	1	0.128	1	1	3≈
	rs	1	1	1	1	0.64	1	1	0.84	1	0.6	1	1	0/
	p	1.01E−6	3.61E−2	1.04E−2	7.90E−3	1.23E−6	6.05E−3	1.16E−8	4.59E−5	4.47E−1	2.90E−1	8.80E−3	1.42E−2	
DRA DE	mean	−13.942	−0.3652	−30660	−6961.8	76.1533	−0.0957	900.255	11787.1	−0.9985	−1.3846	404.503	−5.508	
	std	1.03509	0.08665	3.29605	0.00044	53.9294	0	138.172	1851.41	0.00328	0.1665	97.3722	0	
	rf	0.9824	1	1	1	0.9408	1	1	0.3632	1	0.0416	1	1	
	rs	1	1	1	1	1	1	1	0.84	1	0.68	1	1	

It can be concluded from tables that 1) DRADE performs significantly better in seven out twelve problems of CEC2006 and nine out of twelve problems of CEC2010 than GPEEC, 2) DRADE performs significantly better in eight out twelve problems of CEC2006 and nine out of twelve problems of CEC2010 than KTLBO, 3) DRADE performs significantly better in eight out twelve problems of CEC2006 and six out of twelve problems of CEC2010 than MPMLS, 4) DRADE performs significantly better in eight out twelve problems of CEC2006 and six out of twelve problems of CEC2010 than SACCDE.

Though DRADE is used for distributed and expensive constrained optimization, it still has competitive even better performance than compared SAEAs.

Table 4. Comparison Result of DRADE with Four SAEAs for CEC2010

		c01	c07	c08	c13	c14	c15	c'01	c'07	c'08	c'13	c'14	c'15	
	mean	−0.6085	9.0E+09	9.7E+09	−37.9164	1.6E+14	NaN	−0.16756	9.4E+10	9.3E+10	−1.36175	5.0E+14	4.9E+14	
		(+)	(−)	(−)	(−)	(−)	(/)	(−)	(−)	(−)	(−)	(−)	(≈)	9 −
GPE	std	0.04778	4.0E+09	4.5E+09	4.43071	9.9E+13	NaN	0.01108	1.9E+10	1.5E+10	0	9.9E+13	0	1+
EC	rf	1	1	1	1	0.043	0	1	1	1	0.001	0.102	0.001	1≈
	rs	1	1	1	1	0.96	0	1	1	1	0.04	1	0.04	1/
	p	6.15E−7	1.42E−9	1.42E−9	2.27E−4	2.10E−9	NaN	1.99E−9	1.42E−9	1.42E−9	1.17E−2	1.22E−9	8.00E−1	
	mean	−0.3911	1.0E+08	1.2E+08	−22.281	1.3E+14	6.1E+13	−0.2846	5.3E+09	1.8E+09	−6.88964	2.9E+14	5.0E+14	
		(−)	(−)	(−)	(−)	(−)	(≈)	(+)	(−)	(−)	(−)	(−)	(−)	9 −
KTL	std	0.05844	1.5E+08	1.3E+08	8.72321	8.9E+13	5.3E+13	0.03434	4.4E+09	2.9E+08	6.56405	1.2E+14	2.7E+14	1+
BO	rf	1	1	1	1	0.66667	0.02667	1	1	1	1	0.84	0.06667	2≈
	rs	1	1	1	1	0.92	0.08	1	1	1	1	0.92	0.16	0/
	p	4.60E−3	1.42E−9	1.42E−9	6.57E−9	4.06E−9		1.17E−3	1.42E−9	1.42E−9	3.04E−8	3.16E−9	4.89E−01	
	mean	−0.3947	2.4E+04	5.4E+05	−43.0983	2.7E+13	3.7E+13	−0.22556	4.2E+08	1.3E+09	−15.6600	7.9E+13	3.3E+14	
		(−)	(+)	(≈)	(−)	(≈)	(≈)	(−)	(−)	(−)	(−)	(+)	(≈)	6 −
MP	std	0.07230	1.7E+04	5.6E+05	6.30885	7.9E+13	0	0.03735	5.6E+08	9.5E+08	6.69934	5.7E+13	1.7E+14	2+
MLS	rf	0.86333	0.99667	0.96	0.70267	0.052	0.00067	0.938	0.95667	0.92667	0.132	0.04867	0.00013	4≈
	rs	1	1	1	1	0.92	0.04	1	1	1	0.96	1	0.08	0/
	p	6.60E−3	1.42E−9	1.51E−1	1.70E−2	2.65E−1	5.00E−1	2.83E−2	1.42E−9	1.46E−8	4.85E−7	1.65E−6	2.67E−1	
	mean	−0.3147	3.1E+07	4.7E+07	−51.3691	4.2E+13	NaN	−0.25806	4.5E+08	2.8E+10	−26.2345	2.9E+14	4.4E+14	
		(−)	(−)	(−)	(≈)	(+)	(/)	(≈)	(−)	(−)	(+)	(−)	(≈)	6 −
SAC	std	0.06649	2.7E+07	4.8E+07	7.18763	3.3E+13	NaN	0.05302	1.7E+08	2.1E+10	6.33105	1.2E+14	0	2+
CDE	rf	1	1	1	1	0.10933	0	1	1	1	0.98133	0.16533	0.00267	3≈
	rs	1	1	1	1	0.8	0	1	1	1	1	1	0.04	1/
	p	1.99E−6	1.42E−9	1.42E−9	2.69E−1	5.34E−7	NaN	4.04E−1	1.42E−9	1.42E−9	8.19E−4	3.16E−9	8.00E−1	
	mean	−0.4630	1.0E+06	1.8E+07	−46.7456	6.4E+13	1.8E+14	−0.23281	6.1E+07	7.7E+08	−20.66	2.3E+14	5.5E+14	
DRA	std	0.10171	1.4E+06	1.0E+07	8.9904	4.4E+13	2.3E+13	0.03115	2.1E+08	6.3E+08	10.2707	9.3E+13	0	
DE	rf	1	1	1	1	0.0664	0.0024	1	1	1	0.1522	0.1248	0.0008	
	rs	1	1	1	1	0.88	0.08	1	1	1	0.28	1	0.04	

With the cooperation of the master and work agents, RBF management can effectively assist the expensive evaluation and DE evolution has satisfactory convergence for global search. Therefore, DRADE is a useful algorithm for distributed and expensive constrained optimization problems.

4.4   Application on Optimization of Ceramic Formula

To demonstrate the promising applications of DRADE in engineering optimization problems, we applied it to the optimization of ceramic formula.

Firstly, the optimization of ceramic formula is a distributed and expensive constrained optimization problem due to its characteristics. Ceramic is a traditional manufacturing industry and the optimization of ceramic formula greatly depends on experts experience. The evaluation of an ceramic formula is to fire the ceramic product qualifying production standards, which consume lots of manpower, material resources and time. Therefore, the optimization of ceramic formula is an expensive constrained optimization. Further, different factories have developed different formulae for the same product. Due to trade secret, they do not disclose ceramic formulae to other factories. They authorize an institution with higher credibility to cooperate different factories to optimize the formula. Therefore, the optimization of ceramic formula is a distributed and expensive constrained optimization problem.

Secondly, the optimization of ceramic formula can be mathematically formulated as follows.

$$\min f(\boldsymbol{x}) = \sum_{j=1}^{n} p_j \times x_j \tag{9}$$

$$s.t.\ Q(\boldsymbol{x}) <= 0$$

where n is the total number of gradients. x_j is the percentage and p_j is the unit price of the jth gradient respectively. $Q(\boldsymbol{x})$ is a set of chemical and physical constraints which should be satisfied. For example, the combination of n gradients should have qualified chemical composition. Physical properties like whiteness, hardness, strength, etc. should satisfy fired ceramic product should satisfy national production standards. However, these constraints do not have mathematical formulations and evaluated through different tools.

Thirdly, we get historical formulae and corresponding constrained values of a specific ceramic product $\mathcal{P}$ from an association company, which collect data from different factories. The unit cost of the best qualified formula is 395.21. It can be found in Table 5 that except GPEEC and KTLBO, other three SAEAs can reduce the unit cost than 395.21. Particularly, DRADE has significantly better result than MPMLS and SACCDE.

Therefore, DRADE is an effective method for distributed and expensive constrained optimization and has promising applications in both academia and industry.

Table 5. Comparison Result of DRADE with Four SAEAs for Optimization of Ceramic Formula

	mean	std	rf	rs	p-value
GPEEC	NaN	NaN	0	0	/
KTLBO	NaN	NaN	0	0	/
MPMLS	352.8520	25.6206	0.0493	0.64	7.63E−3
SACCDE	345.9644	32.3060	0.8026	0.96	8.10E−3
DRADE	**326.7871**	32.0233	1	1	

5   Conclusion

In this paper, we propose a distributed RBF-assisted differential evolution algorithm, which is named as DRADE to solve DECOPs. The implementation of DRADE is based on master-slave model. The master agent is responsible for storing expensively evaluated individuals and generating offspring with composite DEs. The distributed work agents are responsible for expensively evaluation and the management of RBF models. In each iteration, the best candidate predicted by RBF models is sent to all work agents for expensive evaluations and stored in all datasets for further training. Experiments on CEC2006, CEC2010 benchmark test suites and the optimization of ceramic formula in engineering demonstrate the promising applications of DRADE.

For the future research, we are going to develop techniques to handle problems with multi objectives and discrete variables based on DRADE. It is meaningful to develop such a new method since many engineering optimization problems usually have more than one objective and their decision variables are partially or even fully discrete.

References

1. Cao, W., Mecrow, B.C., Atkinson, G.J., Bennett, J.W., Atkinson, D.J.: Overview of electric motor technologies used for more electric aircraft (MEA). IEEE Trans. Industr. Electron. **59**(9), 3523–3531 (2012). https://doi.org/10.1109/TIE.2011. 2165453
2. Dong, H., Wang, P., Fu, C., Song, B.: Kriging-assisted teaching-learning-based optimization (KTLBO) to solve computationally expensive constrained problems. Inf. Sci. **556**, 404–435 (2021)
3. Gaetano, G.D., Mundo, D., Maletta, C., Kroiss, M., Cremers, L.: Multi-objective optimization of a vehicle body by combining gradient-based methods and vehicle concept modelling. Case Stud. Mech. Syst. Signal Processing **1**, 1–7 (2015). https://doi.org/10.1016/j.csmssp.2015.06.002, https://www.sciencedirect. com/science/article/pii/S2351988615300026
4. Gong, Y.J., et al.: Automated team assembly in mobile games: a data-driven evolutionary approach using a deep learning surrogate. IEEE Trans. Games, 1–1 (2022). https://doi.org/10.1109/TG.2022.3145886
5. Guo, X., Zhou, M., Liu, S., Qi, L.: Lexicographic multiobjective scatter search for the optimization of sequence-dependent selective disassembly subject to multiresource constraints. IEEE Trans. Cybern. **50**(7), 3307–3317 (2020). https://doi. org/10.1109/TCYB.2019.2901834
6. Handoko, S.D., Kwoh, C.K., Ong, Y.S.: Feasibility structure modeling: an effective chaperone for constrained memetic algorithms. IEEE Trans. Evol. Comput. **14**(5), 740–758 (2010). https://doi.org/10.1109/TEVC.2009.2039141
7. Ibrahim, I., Silva, R., Mohammadi, M.H., Ghorbanian, V., Lowther, D.A.: Surrogate-based acoustic noise prediction of electric motors. IEEE Trans. Magn. **56**(2), 1–4 (2020). https://doi.org/10.1109/TMAG.2019.2945407
8. Ji, J.Y., Yu, W.J., Zhong, J., Zhang, J.: Density-enhanced multiobjective evolutionary approach for power economic dispatch problems. IEEE Trans. Syst. Man, Cybern. Syst. **51**(4), 2054–2067 (2021). https://doi.org/10.1109/TSMC. 2019.2953336
9. Jin, Y., Wang, H., Chugh, T., Guo, D., Miettinen, K.: Data-driven evolutionary optimization: an overview and case studies. IEEE Trans. Evol. Comput. **23**(3), 442–458 (2019). https://doi.org/10.1109/TEVC.2018.2869001
10. Kumar, A., Das, S., Mallipeddi, R.: A reference vector-based simplified covariance matrix adaptation evolution strategy for constrained global optimization. IEEE Trans. Cybern. **52**(5), 3696–3709 (2022). https://doi.org/10.1109/TCYB. 2020.3013950
11. Li, G., Zhang, Q.: Multiple penalties and multiple local surrogates for expensive constrained optimization. IEEE Trans. Evol. Comput. **25**(4), 769–778 (2021). https://doi.org/10.1109/TEVC.2021.3066606

12. Li, J.Y., Zhan, Z.H., Wang, H., Zhang, J.: Data-driven evolutionary algorithm with perturbation-based ensemble surrogates. IEEE Trans. Cybern. **51**(8), 3925–3937 (2021). https://doi.org/10.1109/TCYB.2020.3008280
13. Liang, J.J., et al.: Problem definitions and evaluation criteria for the CEC 2006 special session on constrained real-parameter optimization. J. Appl. Mech. **41**(8), 8–31 (2006)
14. Liu, B., Sun, N., Zhang, Q., Grout, V., Gielen, G.: A surrogate model assisted evolutionary algorithm for computationally expensive design optimization problems with discrete variables. In: 2016 IEEE Congress on Evolutionary Computation (CEC), pp. 1650–1657. IEEE (2016)
15. Liu, B., Zhang, Q., Gielen, G.G.E.: A gaussian process surrogate model assisted evolutionary algorithm for medium scale expensive optimization problems. IEEE Trans. Evol. Comput. **18**(2), 180–192 (2014). https://doi.org/10.1109/TEVC.2013.2248012
16. Mallipeddi, R., Suganthan, P.N.: Problem definitions and evaluation criteria for the CEC 2010 competition on constrained real-parameter optimization. Nanyang Technol. Univ. Singapore , 24 (2010)
17. Peng, C., Liu, H.L., Goodman, E.D.: A cooperative evolutionary framework based on an improved version of directed weight vectors for constrained multiobjective optimization with deceptive constraints. IEEE Trans. Cybern. **51**(11), 5546–5558 (2021). https://doi.org/10.1109/TCYB.2020.2998038
18. Rahi, K.H., Singh, H.K., Ray, T.: Partial evaluation strategies for expensive evolutionary constrained optimization. IEEE Trans. Evol. Comput. **25**(6), 1103–1117 (2021). https://doi.org/10.1109/TEVC.2021.3078486
19. Regis, R.G.: Evolutionary programming for high-dimensional constrained expensive black-box optimization using radial basis functions. IEEE Trans. Evol. Comput. **18**(3), 326–347 (2014). https://doi.org/10.1109/TEVC.2013.2262111
20. Su, Y., Xu, L., Goodman, E.D.: Hybrid surrogate-based constrained optimization with a new constraint-handling method. IEEE Trans. Cybern. **52**(6), 5394–5407 (2022). https://doi.org/10.1109/TCYB.2020.3031620
21. Sun, C., Jin, Y., Cheng, R., Ding, J., Zeng, J.: Surrogate-assisted cooperative swarm optimization of high-dimensional expensive problems. IEEE Trans. Evol. Comput. **21**(4), 644–660 (2017). https://doi.org/10.1109/TEVC.2017.2675628
22. Tian, Y., Zhang, Y., Su, Y., Zhang, X., Tan, K.C., Jin, Y.: Balancing objective optimization and constraint satisfaction in constrained evolutionary multiobjective optimization. IEEE Trans. Cybern. **52**(9), 9559–9572 (2022). https://doi.org/10.1109/TCYB.2020.3021138
23. Wang, H., Jin, Y.: A random forest-assisted evolutionary algorithm for data-driven constrained multiobjective combinatorial optimization of trauma systems. IEEE Trans. Cybern. **50**(2), 536–549 (2020). https://doi.org/10.1109/TCYB.2018.2869674
24. Wang, Y., Yin, D.Q., Yang, S., Sun, G.: Global and local surrogate-assisted differential evolution for expensive constrained optimization problems with inequality constraints. IEEE Trans. Cybern. **49**(5), 1642–1656 (2019). https://doi.org/10.1109/TCYB.2018.2809430
25. Wei, F.F., et al.: A classifier-assisted level-based learning swarm optimizer for expensive optimization. IEEE Trans. Evol. Comput. **25**(2), 219–233 (2021). https://doi.org/10.1109/TEVC.2020.3017865
26. Yang, Z., Qiu, H., Gao, L., Cai, X., Jiang, C., Chen, L.: Surrogate-assisted classification-collaboration differential evolution for expensive constrained optimization problems. Inf. Sci. **508**, 50–63 (2020)

A Flexi Partner Selection Model for the Emergence of Cooperation in N-person Social Dilemmas

Tu Gu and Bo An[(✉)]

School of Computer Science and Engineering,
Nanyang Technological University, Singapore, Singapore
gutu0001@e.ntu.edu.sg, boan@ntu.edu.sg

Abstract. There has been extensive research on social dilemmas. Many models and mechanisms have been proposed to promote cooperation. In this work, we propose a three-stage social dilemma game, the Flexi Partner Selection (FPS) mechanism that can promote cooperative behaviour among agents that are trained to maximize an absolutely selfish objective function. Compared with previous works, our settings are more general and flexible as the number of players in each game is not fixed. Specifically, agents can vote out players based on their past behaviours or stay out of the game if playing the game makes them worse off. Moreover, we consider social dilemmas with both linear and non-linear payoffs. Using reinforcement learning (RL), self-interested agents are able to learn to punish defectors by consistently excluding them and cooperate with others in a number of different settings.

1 Introduction

There has been extensive research on cooperation among self-interested agents in the domains of economics [4] and evolutionary biology [10]. In these contexts, agents usually face the dilemma of choosing between maximizing their individual benefits and cooperating for the sake of the collective good. Studying cooperation in social dilemmas is significant because it is important to learn how to engineer agents and incentive or punishment schemes that enable cooperation to emerge so as to achieve socially desirable outcomes that benefit all [11].

Many mechanisms have been proposed to promote cooperation among self-interested agents in social dilemma games [1,7,13,14]. However, existing methods have several limitations. Firstly, many mechanisms deal with two-person social dilemmas which make them highly inapplicable in more complicated settings like N-person social dilemmas. Secondly, existing mechanisms are proven to work only for very specific social dilemma games. In particular, none of the proposed methods can handle N-person social dilemmas with both linear and non-linear payoffs. Thirdly, irrational players who may take unforeseeable and random actions are generally not taken into consideration. Mechanisms that rely

M. Yokoo et al. (Eds.): DAI 2022, LNAI 13824, pp. 15–28, 2023.
https://doi.org/10.1007/978-3-031-25549-6_2

heavily on actions or outcomes in previous games [1,14] have the risk of breaking down when dealing with irrational players. Last but not least, the number of players in each game is usually fixed which may be a strong and unrealistic assumption in real world contexts. For example, the Collective Risk Dilemma (CRD) [14] which is one of the dilemma games considered in this paper is inspired from climate negotiations. It is not reasonable to assume that the number of participating countries is fixed because some may enter or exit the negotiation over time. As a result, it is important to have a mechanism that remains effective in promoting cooperation even as the number of players in a dilemma game changes.

To address the limitations of existing methods, we propose a three-stage game, the Flexi Partner Selection (FPS) mechanism that allows self-interested agents to learn to cooperate in social dilemmas with both linear and non-linear payoffs. Note that we use a setting that is closer to real world contexts than previous works, i.e., not to fix the number of players in each game and to include irrational players that take unpredictable actions. Inspired by the effectiveness of partner selection [1,14], we first have a voting stage (S1) where agents vote to decide who will be their partners in the dilemma game based on the outcome of the previous game [14]. Then in the opting out stage (S2), agents can opt out if their expected payoffs are negative. In the final stage (S3), players who are still in the game decide whether to cooperate or to defect. In order for the agents to learn from their experiences to maximize their selfish objective functions, they are trained using reinforcement learning (RL). Specifically, agents are trained using Q-Learning [21] in S1 and Deep-Q-Network (DQN) Learning [9] in S3.

With our proposed mechanism, agents learn to consistently exclude "bad" individuals and cooperate with the "good" ones. However, existing partner selection models [1,14] are unable to achieve large scale cooperation when irrational players are involved and the number of players in each game is large and not fixed.

The contributions we make in this work are summarized as follows. Firstly, we propose a three-stage social dilemma game, the Flexi Partner Selection (FPS) mechanism that is able to achieve sustainable cooperation among self-interested agents in social dilemmas with both linear and non-linear payoffs. We also take irrational players into consideration and do not fix the number of players in each game which we believe is closer to real world contexts. Secondly, we design state features and reward functions to facilitate the effective learning of RL agents. Thirdly, we show the effectiveness of our mechanism using a number of different experimental settings. Our experimental results show that self-interested agents are largely able to learn to consistently exclude defectors and cooperate with others. Note that cooperation is not achieved by making the agents sacrifice their individual benefits. Instead, cooperation benefits them in the long run.

2 Related Work

Social dilemmas include 2-person social dilemmas and N-person social dilemmas.

One classic example of a 2-person social dilemma is the 2-person Prisoner's Dilemma where each player can maximize his payoff by defecting while total gains can be maximized when both players cooperate. A partner selection model was proposed in [1] to promote cooperation in repeated 2-person Prisoner's Dilemma through partner selection based on actions in the previous game. Eventually more than 90% of the players learn to choose partners who cooperated in the previous game and then to cooperate with them. However, this model is restricted to 2-person settings.

There are also other works that explore cooperation in 2-person social dilemmas. For example, multi-agent reinforcement learning (MARL) is used to train players to cooperate in computer games such as Wolfpack [6] and the Apple-Pear game [20]. In [12], cooperation is achieved based on the assumption that agents are emotional and hence have guilt when defecting. Under this assumption, agents are not entirely self-interested. The mechanism proposed in [13] assumes that players' past actions are not public and can be revealed by their co-players at a cost.

In the case of N-person social dilemmas, one typical example is the Collective Risk Dilemmas (CRD) which have non-linear payoffs [8,14,15,17]. In a CRD game, each agent can cooperate by contributing a certain amount towards the collective goal or defect by not contributing anything. The collective goal is achieved if the proportion of cooperators is no less than a certain threshold, then every agent receives a positive payoff. Otherwise every agent receives a payoff of 0. Defection is the dominant action although cooperation is desirable. A partner selection model was proposed in [14] in the context of CRD. Agents largely learn to pick up the Outcome-based Cooperative Strategy (OC), i.e., to always cooperate and to select agents that cooperated in the previous game as partners if the collective goal was not achieved in the previous game. One main disadvantage of the proposed model is that agents tend to defect when (1) the amount of contribution required for cooperation is large; and (2) the threshold to achieve the collective goal is large. There are also other ways to punish defectors or to reward cooperators [2,3,19]. However, these mechanisms involve components that incur extra costs such as costly monitoring institutions.

An extension of the CRD is the repeated Public Goods Game (PGG) where each agent can decide how much to contribute and the collective goal is achieved if the total contributed amount is no less than a certain threshold [7]. Although large scale cooperation is achieved, agents fail to learn to cooperate unless a significant portion of the agents first cooperate. There also exist N-person social dilemmas with linear payoffs (LPSD) [16]. In this case, the payoffs to both cooperators and defectors follow some linear functions of the proportion of cooperators. Payoffs to defectors are always larger than that to cooperators. However, the payoff to each cooperator when every agent cooperates is larger than the payoff to each defector when every agent defects. Note that no previous work has achieved large scale cooperation in this setting.

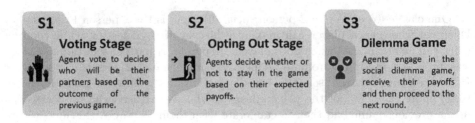

Fig. 1. Flexi partner selection (FPS) mechanism

3   Flexi Partner Selection (FPS) Mechanism

In this section, we formally introduce our proposed three-stage social dilemma game, the Flexi Partner Selection (FPS) mechanism. We consider N players playing rounds of games. Among the N players, we first follow [14] and have N_c players that always cooperate, N_d players that always defect and N_{rl} agents that are to learn a policy in order to achieve the highest possible individual reward. On top on that, we also have N_r irrational players that cooperate (and defect) with a probability of 50%. Note that the values of N, N_c, N_d, N_r and N_{rl} are not fixed and different values will be tested to show the robustness of our proposed method. We aim to make the agents learn to balance the immediate rewards for defecting with the future cost of being excluded and to handle different types of players. An overview of our proposed mechanism is shown in Fig. 1.

3.1   Mechanism Flow

Voting Stage (S1). At the beginning of every round, each of the N_{rl} agents can vote out certain players. There are in total 4 types of votes including: (1) not to vote out anyone; (2) to vote out players who defected in the previous game; (3) to vote out players who defected or got voted out in the previous game; and (4) to vote out players who cooperated in the previous game. Note that although it is counter-intuitive to vote out cooperators, it is important to include this possibility to show that the agents can learn to filter out bad actions. If no more than half of the agents cast the same vote, nobody will be voted out. Note that agents can keep excluding defectors which is fundamentally different from strategies that are entirely based on players' actions in the previous game [1,14].

Opting Out Stage (S2). Based on the outcome of S1, an agent will opt out and receive a payoff of 0 if the expected payoff is negative. The estimation of the expected payoff is done based on the assumption that players who are not voted out in S1 will take the same action as they did when they last participated in the game. This assumption is consistent with existing partner selection models [1,14].

Dilemma Game (S3). All remaining players then decide whether to cooperate or to defect. We consider 2 different social dilemma games, i.e., Social Dilemmas with Linear Payoffs (LPSD) [16] and Collective Risk Dilemmas (CRD) [14]. For LPSD games, the payoffs to a cooperator and a defector are $C(x) = a_c x + b_c$ and $D(x) = a_d x + b_d$ respectively where $x = \frac{m}{N}$, m is the number of cooperators and N is the total number of players. Note that the following 2 conditions have to be satisfied: (1) each player always receives a higher payoff for defecting than for cooperating, i.e., $D(\frac{m}{N}) > C(\frac{m+1}{N})$; and (2) each player receives a lower payoff if all defect than if all cooperate, i.e., $D(0) < C(1)$. For CRD games, each player can cooperate by contributing an amount c towards the collective goal or defect by contributing nothing. The collective goal is achieved if the proportion of cooperators is no less than a certain threshold *thres*. Each player receives a payoff of b (where $b > c$) if the collective goal is achieved and a payoff of 0 otherwise. In both games, defection is the dominant action although cooperation is desirable.

3.2   Policy Learning Using Reinforcement Learning

RL is a suitable tool to train agents in social dilemmas mainly because of 2 reasons. Firstly, RL does not assume any knowledge of the environmental dynamics and directly trains agents to take actions based on the states observed. It can therefore be applied even when the environment is complex. Secondly, RL agents are designed to take actions to maximize their rewards which are usually derived from selfish objective functions for social dilemmas. This design aligns with the characteristic of self-interested agents. It is rather challenging to design the reward system because the actual return an agent receives for his action is not only the payoff of the current game but also reflected in what other agents learn. For example, when an agent defects in a game, he may be punished by other agents and have to bear the consequences in subsequent games. Taking this feature into consideration, the reward received by an agent is designed to contain both the payoff from the current game and the payoffs from subsequent games.

RL Agent Training with Q-Learning in S1. In Q-Learning [21], an agent is to learn a policy represented through a state-action value function $Q(s, a)$. A Q-Table is commonly used in this context. The ϵ-greedy policy is defined as:

$$\pi(s) = \begin{cases} \arg \max_a Q(s, a) & \text{with probability } 1 - \epsilon \\ \text{random action } a & \text{with probability } \epsilon \end{cases}.$$

Each agent stores a set of transitions (s, a, r, s') by interacting with the environment and updating its policy using

$$Q(s, a) \leftarrow Q(s, a) + \alpha[r + \gamma \max_{a'} Q(s', a') - Q(s, a)] \tag{1}$$

where s is the current state, a is the current action, r is the reward received for (s, a), s' is the next state, α is the learning rate and γ is the discount factor. In order to ensure that all (s, a, r, s') tuples stored in each agent's memory buffer are always the most recent and updated, the memory buffers are refreshed whenever a new (s, a, r, s') is available. Each agent can then be trained with the most recent experiences.

In S1, each RL agent is to pick an action $a_v \in \{0, 1, 2, 3\}$. Specifically, $a_v = 0$ is to vote nobody out, $a_v = 1$ ($a_v = 3$) is to vote out every player that defected (cooperated) in the previous game and $a_v = 2$ is to vote out every player that defected or got voted out in the previous game. An agent receives a reward r_v in S1 only if its vote is consistent with the outcome of the voting stage. The reward r_v is the agent's payoff from the dilemma game. Motivated by the OC strategy [14], each RL agent makes the voting decision a_v based on the number of cooperators (N_c^{prev}) and defectors (N_d^{prev}) in the previous game respectively which form the state s_v each RL agent observes.

Each agent learns an optimal policy π_v to take action a_v given state s_v using a Q-Table of size $(N + 1)^2 \times 4$. The corresponding Q-value for the state $s_v = (N_c^{prev}, N_d^{prev})$ and action a_v is found at the $(N \cdot N_c^{prev} + N_d^{prev})^{th}$ row and a_v^{th} column of the Q-Table. An ϵ-greedy policy is used with initial ϵ_v value, $\epsilon_{v(init)}$ set to be 1 so as to allow more exploration at the start. ϵ_v is reduced by 1% after each round but never allowed to be below 0.1, i.e., $\epsilon_{v(end)} = 0.1$. Discount rate γ and learning rate α are set to be 0.99 and 0.1 respectively.

RL Agent Training with DQN Learning in S3. Deep-Q-Network (DQN) Learning is used in S3 because it is no longer practical to use Q-Learning to trace all state-action pairs [9]. In this context, a Q-value function $Q^*(s, a)$ is used to represent the expected accumulated reward that the agent can obtain if it takes action a in state s and then follows the optimal policy until it reaches a terminal state. The optimal policy takes the action with the maximum Q-value in any state. The Q-value function $Q^*(s, a)$ is approximated with a deep neural network $Q(s, a; \Theta)$ with parameters Θ. We adopt the ϵ-greedy deep Q-learning with experience replay [9] for learning the Q-value functions.

Given a set of transitions (s, a, r, s'), parameters in $Q(s, a; \Theta)$ are updated with a gradient descent step by minimizing the mean square error (MSE) loss function, as shown in Eq. 2.

$$L(\theta) = \sum_{s,a,r,s'} [(r + \gamma \max_{a'} \hat{Q}(s', a'; \Theta^-) - Q(s, a; \Theta))^2], \qquad (2)$$

where γ is the discount rate, and $\hat{Q}(; \Theta^-)$ is the frozen target network.

In S3, inspired by the strategy that relies on players' actions in the previous game proposed in [1], each RL agent takes an action $a_d \in \{$Cooperation, Defection$\}$ based on the state s_d observed which includes relevant information about players that participate in the current game. The state consists of the proportion of players that cooperated, defected, got voted out and opted out in the previous game respectively. Intuitively, the reward r_d each agent receives can simply be the payoff from the social dilemma game. However, because of the presence of S1, one's action in S3 may have an impact on whether or not he will be voted out in subsequent games. As a result, our design of the reward signal follows other RL applications [18,22] and consists of both immediate payoff and short-term

Table 1. Emergence of cooperation in Different Mechanisms

	PSM1	PSM2	FPS
Avg% of Cooperators	65%	62%	95%
Avg Individual Payoff	2.05	1.94	2.41

future payoffs. Specifically, the reward r_d is set to be the sum of the payoff from the social dilemma game in the current round and the payoffs from the next 2 rounds using a discount factor of 0.9.

Each agent learns an optimal policy π_d to take action a_d given state s_d using DQN Learning. We parameterize each RL agent's model using a neural network with one hidden layer of 32 neurons with activation function SELU [5]. Other settings are the same as in S1.

4   Experiments and Discussions

In this section, we evaluate the emergence of cooperation with our proposed FPS mechanism, as described in Sect. 3.1, using two different types of social dilemma games, i.e., Social Dilemmas with Linear Payoffs (LPSD) and Collective Risk Dilemmas (CRD). We follow [1] and train RL agents in 30,000 rounds of games (episodes). The default settings for our experiments are as follows:

- **LPSD**: We follow [16] and set the payoff functions as $C(x) = a_c x + b_c$ for ($a_c = 2, b_c = 0.5$) and $D(x) = a_d x + b_d$ for ($a_d = 2, b_d = 1$) respectively.
- **CRD**: We follow [14] and set the net gain to a cooperator when the collective goal is achieved to be 2. Specifically, the contributed amount from a cooperator is $c = 1$, the payoff to each player if the collective goal is met is $b = 3$ and the proportion of cooperative behaviour required to achieve the collective goal is $thres = 0.7$.
- **Players**: The number of players that always cooperate is $N_c = 5$, the number of players that always defect is $N_d = 5$, the number of irrational players that take random actions is $N_r = 5$ and the number of RL agents is $N_{rl} = 5$. Under this default setting, the total number of players in a game is significantly more than that in previous works [1,14].

Note that all experiments follow the default setting unless specified otherwise.

4.1   Emergence of Cooperation in FPS Mechanism

In this experiment, we evaluate the effectiveness of the FPS mechanism in promoting cooperation and make comparisons with the partner selection models proposed in [1,14] which we call PSM1 and PSM2 respectively. The optimal strategy of PSM1 is to always choose a cooperator as the partner and cooperate, and to cooperate (defect) when chosen by one who cooperated (defected) in the previous game. The optimal strategy of PSM2 is to always cooperate, and to only choose cooperators as partners if the collective goal was not achieved in the previous game. We make slight modifications to PSM1 and PSM2 so as to make them fit into our setting where the number of players per game is not fixed. For PSM1, we modify it as a mechanism where agents choose all players that cooperated in the previous game and then cooperate only if in the previous game all players cooperated (for LPSD) or the collective goal was achieved (for CRD). For PSM2, we modify it as a mechanism where if in the previous game not all players cooperated (for LPSD) or the collective goal was not achieved (for CRD), agents would choose players who cooperated in the previous game to be their partners and cooperate. Otherwise, they would choose all other players to be their partners and cooperate.

For each mechanism, the games are repeatedly played 100 times using the default setting and its respective optimal strategy. We then report the average proportion of cooperative behaviour among S3 participants and the average payoff each agent receives in Table 1. We observe that PSM1 and PSM2 show rather poor performance as they achieve an average of 65% and 62% cooperation respectively. In contrast, the FPS mechanism is able to achieve 95% cooperation. We observe that the RL agents demonstrate significantly better cooperative behaviour than the agents adopting respective optimal strategies in PSM1 and PSM2. Moreover, the RL agents in the FPS mechanism receive the best payoffs which shows that cooperation is not achieved at the expense of their individual payoffs. The poor performance of PSM1 and PSM2 may be attributed to 3 main reasons. Firstly, both mechanisms are not proven to be effective for social dilemmas with more than 7 players. Their applications may be highly restricted to small groups. Secondly, as the number of players in each game is not fixed, they may not adapt well enough to such a complicated environment. Thirdly, as they rely heavily on players' actions and games outcomes in previous games, they are not able to effectively deal with irrational players whose actions are unpredictable.

4.2 Learned Strategy of RL Agents

In this set of experiments, we aim to explore how the behaviour of RL agents and cooperative behaviour evolve in S1 and S3 respectively. We follow [1] and report the evolvement of each action in Figs. 2 and 3 respectively. Specifically, we report the average proportion of each action per 100 rounds (episodes) of games. Note that there is no exploration in the last 100 rounds of games, i.e., all RL agents make decisions based on their learned policies.

RL Agents' Learned behaviour in S1. Based on the result shown in Fig. 2, we highlight 3 key observations.

Firstly, the average proportion of RL agents taking action $a_v = 2$ in the last 100 episodes is 83% for LPSD and 91% for CRD. In both games, most RL agents learned to vote out players that defected or got voted out in the previous game as they recognize that in order to maximize their individual payoffs, it is insufficient to only exclude players that defected in the previous game like

the mechanisms designed in [1,14]. Instead, players that have been punished for
defecting must be consistently excluded.

Secondly, in the first few thousand episodes, the number of RL agents taking
$a_v = 1$ increased before most of them eventually learned to choose $a_v = 2$. This
is mainly because $a_v = 1$ is also effective in punishing defectors. However, over
time, most RL agents learned that $a_v = 2$ is the most effective as they try to
maximize their individual payoffs.

Last but not least, we observe that RL agents take action $a_v = 2$ more
often in CRD games than in LPSD games. Note that CRD games have only 2
possible outcomes, i.e., whether or not the collective goal is achieved. In contrast, LPSD games have different payoffs for different proportions of cooperative
behaviour. As LPSD games are inherently more complicated than CRD games,
it is expectedly harder for RL agents to learn an optimal policy in LPSD games.
It also explains why it takes a longer time for RL agents to learn not to vote out
cooperators $(a_v = 3)$ in LPSD games.

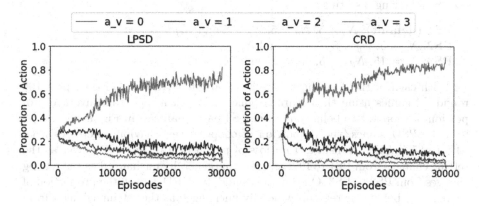

Fig. 2. Evolvement of behaviours of RL agents in S1

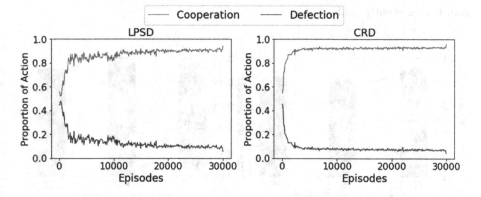

Fig. 3. Emergence of cooperation in S3

Emergence of Cooperation in S3. As shown in Fig. 3, the average proportion of cooperative behaviour among S3 participants in the last 100 episodes is 94% for LPSD and 96% for CRD. Similar to S1, we also observe in S3 that it takes a longer time for convergence to occur in LPSD games than in CRD games. Note that although LPSD games are more complicated than CRD games as explained earlier, there is only marginal difference in the proportion of cooperators between these 2 types of games. It shows that the FPS mechanism remains robust even as the payoff type of the game changes.

4.3   The Effect of Changing the Number of Players

In this experiment, we aim to evaluate the robustness of the FPS mechanism as the number of players increases. As more RL agents explore and adjust their policies, there is a larger shift in the dynamics of the environment which may affect the efficiency and the effectiveness of their learning. Therefore, we run experiments with up to 3 times as many players in each game. Specifically, we use the following 3 settings:

- **PS1 (default)**: $N_c = 5$, $N_d = 5$, $N_r = 5$ and $N_{rl} = 5$.
- **PS2**: $N_c = 5$, $N_d = 5$, $N_r = 5$ and $N_{rl} = 15$.
- **PS3**: $N_c = 15$, $N_d = 15$, $N_r = 15$ and $N_{rl} = 15$.

With each of the 3 settings, RL agents are first trained and then play 100 rounds of games using their learned policies. We then report the average proportion of cooperative behaviour among S3 participants in in Fig. 4.

For LPSD games, the proportion of cooperative behaviour decreases from 94% to 85% which translates to a loss of payoff of 0.18 to every player as the setting changes from PS1 to PS2 and remains generally unchanged as the setting changes from PS2 to PS3. On the other hand, for CRD games, the proportion of cooperative behaviour remains generally unchanged as the setting changes from PS1 to PS2 and decreases from 96% to 91% as the setting changes from PS2 to PS3. Note that the outcomes of CRD games are not affected as the collective goal is consistently achieved.

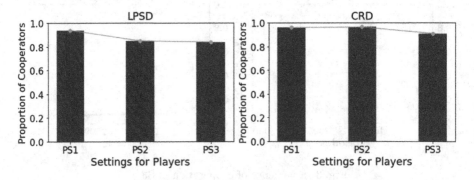

Fig. 4. Impact of varying the number of players on the emergence of cooperation

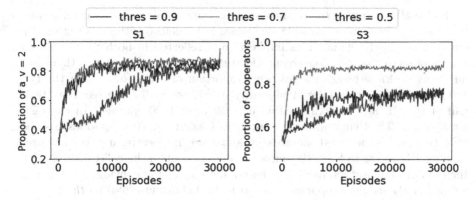

Fig. 5. Impact of varying *thres* for CRD games

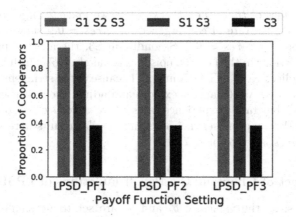

Fig. 6. Impact of varying the payoff functions for LPSD games

We have shown that, the FPS mechanism remains consistently effective as the number of RL agents increases from 5 to 15 and the total number of players increases from 20 to 60.

4.4 The Effect of Changing the Collective Goal Requirement of CRD Games

We have shown experimentally the emergence of cooperation in CRD games with the default setting of *thres* = 0.7. Any change in *thres* is expected to have a large impact on the dynamics of the environment and hence RL agents' learning. In Fig. 5, we report the average proportion of action $a_v = 2$ in S1 and the average proportion of cooperative behaviour in S3 per 100 rounds of game for *thres* $\in \{0.5, 0.7, 0.9\}$ [14]. Note that there is no exploration in the last 100 rounds of game.

For smaller values of *thres*, there should be more tolerance toward defectors because it is now more possible for someone to defect without affecting the outcome of a CRD game. This implication is reflected in both S1 and S3. In S1, we observe that the convergence for *thres* = 0.5 is much slower than that for *thres* = 0.7 although the average proportion of action $a_v = 2$ in the last 100 episodes does not differ significantly. For *thres* = 0.7, we see more than half of the RL agents taking action $a_v = 2$ after 1000 episodes. In contrast, for *thres* = 0.5, that only occurs after no fewer than 7000 episodes. As the collective goal is achieved more easily and more frequently, it takes a longer time for RL agents to learn the impact of defection upon the collective goal and then to punish defectors. In S3, we observe not only slower convergence, but also a significantly smaller proportion of cooperative behaviour (76%) for *thres* = 0.5 than that (96%) for *thres* = 0.7. Note that the outcomes of the CRD games are not affected.

For larger values of *thres*, i.e., *thres* = 0.9, we make 2 major observations. First of all, in S1, the rate of convergence for *thres* = 0.9 shows no difference compared to that for *thres* = 0.7. Secondly, in S3, the average proportion of cooperative behaviour in the last 100 episodes is merely 78% which is not enough to achieve the collective goal. This is mainly because of the stringent requirement for collective success. As RL agents explore and adjust their policies during the learning process, they rarely experience cases where the collective goal is achieved which significantly compromises their learning. Similar results were also reported in [14] where *thres* is set to be 6/7.

4.5   The Effect of Changing the Payoff Functions of LPSD Games

In LPSD games, as the values of b_c and b_d are set to be smaller and more negative, it may affect RL agents' decisions in each stage and also the dynamics of the environment because a larger proportion of cooperators will be needed in order to achieve non-negative payoffs. In this set of experiments, we have 2 objectives: (1) evaluate the robustness of the FPS mechanism amid changes in the payoff functions; and (2) explore the roles of S1 and S2 by first training RL agents without them and then adding them incrementally. Our experiments are run with the following payoff functions:

- **LPSD_PF1**: $a_c = a_d = 2$, $b_c = -0.5$ and $b_d = 0$.
- **LPSD_PF2**: $a_c = a_d = 2$, $b_c = -1$ and $b_d = -0.5$.
- **LPSD_PF3**: $a_c = a_d = 2$, $b_c = -1.5$ and $b_d = -1$.

With each payoff function, RL agents are trained and 100 rounds of games are played. We then report the average proportion of cooperative behaviour among S3 participants in Fig. 6.

First of all, when RL agents are trained with only S3, the proportion of cooperative behaviour stays consistently around 37.5% which consists of players that always cooperate and irrational players that cooperate half the time. The RL agents learn to defect as it is the dominant action. Note that in this case defectors are never punished. Similar results are also reported in [1] in the 2-person setting. As S1 is added, most RL agents learn to consistently exclude players with "bad" behaviour and to cooperate themselves. The proportion of cooperative behaviour in S3 increases by more than 100% to at least 80% for all 3 settings. Furthermore, as S2 is added and RL agents can opt out of the game, the proportion of cooperative behaviour in S3 further increases to at least 91% which then translates to gains of payoffs of 0.2, 0.22 and 0.16 to each player in LPSD_PF1, LPSD_PF2 and LPSD_PF3 respectively.

We can draw 2 conclusions from this experiment. Firstly, the FPS mechanism remains effective amid changes in the payoff functions of LPSD games. Secondly, S1 and S2 significantly help with RL agents' learning and contribute to the emergence of cooperation.

5 Conclusion

In this work, we propose a three-stage N-person social dilemma game, the Flexi Partner Selection (FPS) mechanism to promote cooperation among self-interested agents. The FPS mechanism is experimentally proven to be effective for social dilemmas with both linear and non-linear payoffs without the need to fix the number of players in a game. It also demonstrates advantages over existing methods when dealing with irrational players. In order to facilitate RL agents' learning, we carefully design the state based on which they make decisions and reward functions that provide valuable feedback to their past actions. RL agents largely learn to consistently exclude defectors and to cooperate themselves.

We believe that with this work, we would open a few directions for future work. The first possible direction is to explore possible extension to social dilemmas with more complicated settings such as Public Goods Game (PGG) [7]. The second possible direction is the emergence of cooperation in evolving environments such as payoff functions that change over time. Last but not least, it is possible to explore other machine learning applications in social dilemmas.

References

1. Anastassacos, N., Hailes, S., Musolesi, M.: Partner selection for the emergence of cooperation in multi-agent systems using reinforcement learning. In: Proceedings of the AAAI Conference on Artificial Intelligence, vol. 34, pp. 7047–7054 (2020)
2. Chen, X., Sasaki, T., Brännström, Å., Dieckmann, U.: First carrot, then stick: how the adaptive hybridization of incentives promotes cooperation. J. R. Soc. Interface 12(102), 20140935 (2015)
3. Fehr, E., Gächter, S.: Altruistic punishment in humans. Nature 415(6868), 137–140 (2002)
4. Gintis, H., Bowles, S., Boyd, R.T., Fehr, E., et al.: Moral Sentiments and Material Interests: The Foundations of Cooperation in Economic Life, vol. 6. MIT Press, Cambridge (2005)
5. Klambauer, G., Unterthiner, T., Mayr, A., Hochreiter, S.: Self-normalizing neural networks. In: Proceedings of the 31st International Conference on Neural Information Processing Systems, pp. 972–981 (2017)

6. Leibo, J.Z., Zambaldi, V., Lanctot, M., Marecki, J., Graepel, T.: Multi-agent reinforcement learning in sequential social dilemmas. arXiv preprint. arXiv:1702.03037 (2017)
7. Li, K., Hao, D.: Cooperation enforcement and collusion resistance in repeated public goods games. In: Proceedings of the AAAI Conference on Artificial Intelligence, vol. 33, pp. 2085–2092 (2019)
8. Milinski, M., Sommerfeld, R.D., Krambeck, H.J., Reed, F.A., Marotzke, J.: The collective-risk social dilemma and the prevention of simulated dangerous climate change. Proc. Natl. Acad. Sci. **105**(7), 2291–2294 (2008)
9. Mnih, V., et al.: Human-level control through deep reinforcement learning. Nature **518**(7540), 529–533 (2015)
10. Nowak, M.A.: Evolutionary Dynamics: Exploring the Equations of Life. Harvard University Press, Cambridge (2006)
11. Paiva, A., Santos, F., Santos, F.: Engineering pro-sociality with autonomous agents. In: Proceedings of the AAAI Conference on Artificial Intelligence, vol. 32 (2018)
12. Pereira, L.M., Lenaerts, T., Martinez-Vaquero, L.A., Han, T.A.: Social manifestation of guilt leads to stable cooperation in multi-agent systems. In: Proceedings of the 16th Conference on Autonomous Agents and MultiAgent Systems, pp. 1422–1430 (2017)
13. Santos, F., Pacheco, J., Santos, F.: Social norms of cooperation with costly reputation building. In: Proceedings of the AAAI Conference on Artificial Intelligence, vol. 32 (2018)
14. Santos, F.P., Mascarenhas, S.F., Santos, F.C., Correia, F., Gomes, S., Paiva, A.: Outcome-based partner selection in collective risk dilemmas. In: Proceedings of the 18th International Conference on Autonomous Agents and MultiAgent Systems, pp. 1556–1564 (2019)
15. Santos, F.C., Pacheco, J.M.: Risk of collective failure provides an escape from the tragedy of the commons. Proc. Natl. Acad. Sci. **108**(26), 10421–10425 (2011)
16. Szilagyi, M.N.: An investigation of n-person prisoners' dilemmas. Complex Syst. **14**(2), 155–174 (2003)
17. Tavoni, A., Dannenberg, A., Kallis, G., Löschel, A.: Inequality, communication, and the avoidance of disastrous climate change in a public goods game. Proc. Natl. Acad. Sci. **108**(29), 11825–11829 (2011)
18. Trummer, I., Wang, J., Maram, D., Moseley, S., Jo, S., Antonakakis, J.: Skinnerdb: Regret-bounded query evaluation via reinforcement learning. In: Proceedings of the 2019 ACM SIGMOD International Conference on Management of Data, pp. 1153–1170 (2019)
19. Vasconcelos, V.V., Santos, F.C., Pacheco, J.M.: A bottom-up institutional approach to cooperative governance of risky commons. Nat. Clim. Chang. **3**(9), 797–801 (2013)
20. Wang, W., Hao, J., Wang, Y., Taylor, M.: Towards cooperation in sequential prisoner's dilemmas: a deep multiagent reinforcement learning approach. arXiv preprint. arXiv:1803.00162 (2018)
21. Watkins, C.J., Dayan, P.: Q-learning. Mach. Learn. **8**(3–4), 279–292 (1992). https://doi.org/10.1007/BF00992698
22. Yang, Z., et al.: Qd-tree: learning data layouts for big data analytics. In: Proceedings of the 2020 ACM SIGMOD International Conference on Management of Data, pp. 193–208 (2020)

Efficient Deep Reinforcement Learning via Policy-Extended Successor Feature Approximator

Yining Li[1], Tianpei Yang[1,2(✉)], Jianye Hao[1(✉)], Yan Zheng[1],
and Hongyao Tang[1]

[1] College of Intelligence and Computing, Tianjin University, Tianjin, China
{yiningli,jianye.hao,yanzheng,bluecontra}@tju.edu.cn
[2] University of Alberta, Edmonton, Canada
tpyang@tju.edu.cn

Abstract. Successor Features (SFs) improve the generalization of Reinforcement Learning across unseen tasks by decoupling the dynamics of the environment from the rewards. However, the decomposition highly depends on the policy learned on the task, which may not be optimal in other tasks. To improve the generalization of SFs, in this paper, we propose a novel SFs learning paradigm, Policy-extended Successor Feature Approximator (PeSFA) which decouples the SFs from the policy by learning a policy representation module and inputting the policy representation to SFs. In this way, when we fit SFs well in the policy representation space, we can directly obtain a better SFs corresponding to any task by searching the policy representation space. Experimental results show that PeSFA significantly improves the generalizability of SFs and accelerates the learning process in two representative environments.

Keywords: Reinforcement learning · Transfer learning · Successor features · Policy representation

1 Introduction

Reinforcement Learning (RL) is used to solve sequential decision-making problems, where an agent learns its policy by interacting with the environment [11,18]. Recent advance has shown Deep RL (DRL) obtains expressive success of achieving human-level control in complex tasks [13,17]. However, DRL is still faced with the problem of sample inefficiency. Transfer learning mainly accelerates learning by using prior knowledge, which takes the knowledge gained on source tasks and uses it to learn a different but related target task more efficiently. It is a promising way to improve the sample efficiency of DRL [20–22].

The successor features (SFs) [3] transfer the knowledge of the task-independent policy among tasks with the Generalised Policy Improvement (GPI) [3] method by decoupling the dynamics of the environment from the rewards.

M. Yokoo et al. (Eds.): DAI 2022, LNAI 13824, pp. 29–44, 2023.
https://doi.org/10.1007/978-3-031-25549-6_3

However, SFs has its limitations [3]. It cannot guarantee that the policy obtained by the GPI method is the optimal policy for the new task, so further learning and exploration are needed. Besides, some algorithms apply SFs in various ways to exploit its rapid task inference mechanism. The SFOLS [1] and SIP [2] construct a better basic policy set in different ways and obtain a policy that can perform well on more complex downstream tasks with the GPI method. But for more complex environments, constructing the policy set is computationally intensive. VISR [15] and APS [12] take the policy-conditioning variable as an additional input to SFs and combine the application of mutual information to learn different behavior policies. However, generalizing SFs over behavioral policies implies inefficiencies in sampling and generalization among policies. USFA [4] treats each task's weight as the representation of the optimal policy for the corresponding task, then takes them as additional input to the SFs, allowing SFs to have generalization capability among tasks. However, USFA has the problem of insufficient generalization among tasks and faces the problem of unsmoothness in the task space caused by directly equating task weights with policies. Still, none explicitly exploit the relationship between policies and SFs to allow successor features to generalize among policies and tasks, nor do they address the problem of over-coupling SFs with policies. If we do not decouple SFs from policy, much of the previously available policy information will be forgotten in the process of learning SFs, which makes knowledge transfer inefficient. Moreover, similar policies may be repeatedly learned when learning policies for different tasks, making the learned knowledge redundant.

In this paper, we aim to decouple SFs from policy, which helps us take advantage of generalizing SFs among policies to improve learning efficiency. Therefore, we encode policy into representation as an additional input to the SFs and propose policy-extended successor features approximator (PeSFA). Its role is to fit the SFs corresponding to each policy in the policy representation space. So given a representation of any policy, we can well estimate the SFs of that policy. At the same time, we know that SFs can decouple the policy from the task, then we càn decouple the SFs from the policy to generalize SFs among tasks and policies. With the help of this property, we can find the SFs corresponding to a better policy for any task in the policy representation space. The primary method we use is to search for a better policy corresponding to the given task in the policy representation space in the direction of the gradient that maximizes the action-value function. Experimental results demonstrate the effectiveness of PeSFA in improving learning performance compared with previous methods.

2   Related Work

Extension of SFs Function. UVFA [16] extends the traditional value function by taking additional input and generalizes the value function over different targets by designing additional input forms. USFA [4] takes task weights as additional inputs to SFs, which generalize SFs at the optimal policy level by equating the task weights to the encoding of the corresponding optimal policy. However,

USFA directly equates the optimal policy with weight, making its generalizability among tasks affected by the continuity between tasks and optimal policies, leading to inaccurate estimation. VISR [10] applies SFs to the unsupervised pre-training domain by using policy conditioning variable as additional input to SFs and learning different behavioral policies by maximizing the behavioral mutual information while using the GPI method to accelerate the speed of policy learning. On the other hand, the APS algorithm [12] points out some problems in the VISR and APT algorithms, then combines the two as APS, which solves the issue of the lack of exploration capability of VISR. Both VISR and APS generalize SFs over the policy conditioning variable, making the method inefficient in terms of sampling during training and generalization at the policy level.

Constructing Policy Set. The SIP [2] proposes to construct a specific set of mutually independent policies. By using the GPI method on this policy set, policies with higher performance can be obtained immediately on downstream tasks that are usually more complex. SFOLS [1] enables the GPI method to get the optimal policy for any task without interacting with the environment by constructing a convex coverage policy set. Once the environment becomes complex, SIP and SFOLS will need to build a set containing a large number of policies, which requires a lot of computation. The Policy Caches [14] can calculate the upper bound performance of the policy obtained by the GPI method on a new task with the information in the historical policy set. It can help us determine whether a new policy needs to be learned or uses the policy obtained by the GPI method.

Transfering Expert Experience with SFs. The PsiPhi-Learning [7] algorithm learns the SFs using the ITD method over a series of trajectories with no reward. The trajectories are generated by agents with arbitrary performance interacting with the environment. And GPI method combines the policies obtained by each agent and the current policy being learned to get a policy with better performance. However, this approach requires using many models to fit the SFs under the corresponding policies, which lacks generalizability. Abstraction with Successor Features [9] defines abstract successor options under the SMDP setting. And each possible starting point is calculated to match the abstract successor option under the specified MDP by the feature-matching algorithm in IRL. Then the policy of the abstract successor option under the specified MDP can be obtained to transfer the abstract option in different environments. However, this method essentially transfers knowledge through a feature-matching-based IRL approach, which lacks task-specific generalization.

Our work belongs to the category of extending SFs by adding additional input. First, we decouple SFs from policies by using policy representations as an additional input, which allows it to generalize among policies and leverage the knowledge of historical policies. In addition, with the property of generalizing SFs among policies and tasks, a method is proposed to search for a better policy for the corresponding task in the policy representation space. Besides, the search method can also help us expedite the learning process and explore the policy representation space more efficiently, accelerating the fitting process of SFs and corresponding policies.

3   Background and Problem Formulation

3.1   Reinforcement Learning

Reinforcement learning (RL) means that the agent guides behavior through the reward obtained by interacting with the environment, and the goal is to make the agent obtain the maximum reward. The model of the interaction between the agent and environment can be modeled as a Markovian decision process (MDP).

MDP can be expressed as a tuple $\mathcal{M} = (\mathcal{S}, \mathcal{A}, p, \mathcal{R}, \gamma)$, where $\mathcal{S}$ and $\mathcal{A}$ are the state and action space; $p(\cdot \mid s, a)$ describes the transition dynamics; $\mathcal{R}(s, a, s')$ is the reward obtained in the transition of $s \xrightarrow{a} s'$, and $\gamma \in [0, 1)$ is the discount factor [6].

The goal of the agent in RL is to find a policy π, that can maximize the expected cumulative reward, i.e., $G_t = \sum_{i=0}^{\infty} \gamma^i \mathcal{R}_{t+i+1}$, where $\mathcal{R}_t = \mathcal{R}(S_t, A_t, S_{t+1})$. The action-value function of the policy π can be expressed as

$$Q^\pi(s, a) \equiv \mathbb{E}^\pi(G_t \mid S_t = s, A_t = a) \tag{1}$$

where $\mathbb{E}^\pi[\cdot]$ represents the expected return of following the policy π.

3.2   Successor Features

First, We illustrate the setting of the environment that is of interest in this paper with the MDP tuple. Each task is determined by the reward function R_w, and the elements of the MDP tuple for each task are kept consistent except for R. The one-step expected reward obtained in the transition of $s \xrightarrow{a} s'$ is

$$\mathbb{E}[R_w(s, a, s')] = r_w(s, a, s') = \phi(s, a, s')^\top w \tag{2}$$

where $\phi(s, a, s') \in \mathbb{R}^d$ is feature of the one-step transition (s, a, s') and $w \in \mathbb{R}^d$ is the task weight [3]. The one-step transition feature $\phi(s, a, s')$ can be viewed as a description of a salient activity expected or undesired by the agent, such as picking up an object or walking through a door. Suppose $\phi_t = \phi(s_t, a_t, s_{t+1})$, then the action-value function of Eq. 1 can be written as

$$Q^\pi(s, a) = \mathbb{E}^\pi \left[\sum_{i=t}^{\infty} \gamma^{i-t} \phi_{i+1} \mid S_t = s, A_t = a \right]^\top w = \psi^\pi(s, a)^\top w \tag{3}$$

where $\psi^\pi(s, a)$ is the successor features (SFs) [3], which is the expected cumulative value of ϕ under the policy π. And the following equation can be obtained

$$\psi^\pi(s, a) = \phi_t + \gamma \mathbb{E}^\pi[\psi^\pi(S_{t+1}, \pi(S_{t+1})) \mid S_t = s, A_t = a] \tag{4}$$

This formula shows that SFs satisfy the Bellman equation and ϕ_t can be considered as the reward function in Eq. 1. Therefore, most of the RL methods can be used to calculate ψ^π. With the temporal difference (TD) error [18], the loss function for learning successor features is as follow

$$\mathcal{L}(\tilde{\psi}^{\pi}) = \mathbb{E}_{(s,a,s')\sim\mathcal{D}}\left[\left(\phi(s,a,s') + \gamma\tilde{\psi}^{\pi}(s',a') - \tilde{\psi}^{\pi}(s,a)\right)^2\right] \quad (5)$$

where $a' = \arg\max_b \tilde{Q}^{\pi}(s',b) = \arg\max_b \tilde{\psi}^{\pi}(s',b)$w, $\mathcal{D}$ represents the replay buffer used to store historical data.

Suppose the agent has learned the SFs $\{\psi^{\pi_i}\}_{i=1}^n$ on a series of tasks. Now given a new task w, with the property of the SFs, the action-value function of the existing policy π_i on the new task w can be obtained immediately without the need to evaluate again, i.e., $Q^{\pi_i} = \psi^{\pi_i}(s,a)^{\top}$w. Then the policy on a new task using the GPI method can be defined as

$$\pi(s) \in \arg\max_a \max_i Q^{\pi_i} = \arg\max_a \max_i \psi^{\pi_i}(s,a)^T w \quad (6)$$

And the theorem of GPI guarantees that $\forall(s,a) \in \mathcal{S} \times \mathcal{A}, Q^{\pi}(s,a) \geq \max_i Q^{\pi_i}$. This result can likewise be extended to replace Q^{π_i} with the fitted $\tilde{Q}^{\pi_i}$.

4   Policy-Extended Successor Features

Traditional SFs leverage the knowledge of historical policies to transfer to the new task by obtaining a policy with a basic performance guarantee with the GPI method. However, the policy is not optimal for the new task, and it cannot directly find a better policy built on the sub-optimal policy obtained by the GPI method, but it requires exploration and training. In addition, the learned SFs are highly correlated to the policy, which means that SFs cannot leverage all the valuable policy knowledge learned during the training process of a policy from random to optimal. Besides, it limits the efficiency of SFs in transferring all possible policy knowledge.

In this paper, we propose a novel SFs learning paradigm called Policy-extended Successor Feature Approximator, which takes the policy representation as an additional input to SFs. We can fit the SFs corresponding to the policy representations through the training process, that is, to fit SFs ψ^{π} in the policy representation space Π. And Fig. 1 is the overall architecture diagram of this work. The architecture mainly comprises the SFs module, policy representation module, and gradient ascent based representation space search module.

Ideally, in the policy representation space, if we know any policy representation, the SFs corresponding to the policy can be obtained by inputting the policy representation into the SFs module. To achieve such an effect, we enhance the inter-policy generalization of the SFs module by continuously exploring and fitting the corresponding SFs in the policy representation space. At this time, The gradient ascent based representation space search module can use the gradient direction obtained under the corresponding task as the optimization direction for the representation space searching. When we encounter any new task w, we can randomly select the policy representation as the initial point, then take the gradient as the direction and continuously update the policy representation. This

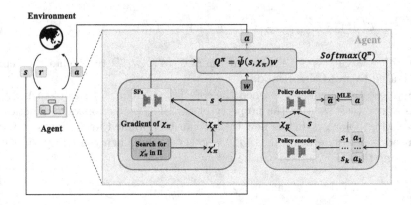

Fig. 1. The overall architecture of PeSFA which consists of SFs module, policy representation module, and gradient ascent based representation space search module.

process allows us to search for a better policy and explore the policy representation space more efficiently. Besides, the policy representation module consists of a policy encoder and an auxiliary decoder which can learn in a self-supervised manner, providing us with the representation encoding method. Next, we will go into the details of each module.

4.1 Policy-Extended Successor Features Approximator

On the basis of UVFA [16], we extend SFs to take policy as an additional input and define the policy-extended successor features as $\psi : \mathcal{S} \times \mathcal{A} \times \Pi \to \mathbb{R}^d$, that is, $\psi(s, a, \pi) = \psi^\pi(s, a)$ for all $s \in \mathcal{S}, a \in \mathcal{A}$ and $\pi \in \Pi$. At which point, the action-value function can be expressed in the following form

$$Q^\pi(s, a, \mathrm{w}) = \psi^\pi(s, a)^\top \mathrm{w} = \psi(s, a, \pi)^\top \mathrm{w} \qquad (7)$$

In practice, using the policy as an additional input requires an algorithm to encode policy into the corresponding representation as a more compact input. Assuming that there is a mapping function $g : \Pi \to \mathcal{X} \subseteq \mathbb{R}^n$, any policy $\pi \in \Pi$ can be mapped into an n-dimensional representation $\chi_\pi = g(\pi) \in \mathcal{X}$. Then Eq. 7 can be expressed as $Q^\pi(s, a, \mathrm{w}) = \tilde{\psi}(s, a, \chi_\pi)^\top \mathrm{w}$, we call $\tilde{\psi}(s, a, \pi) \approx \psi(s, a, \pi)$ a *policy-extended successor features approximator* (PeSFA).

PeSFA decouples the SFs from the policy, and we can obtain the corresponding SFs $\tilde{\psi}(s, a, \chi_\pi)$ according to the representation χ_π of different policy π. It is equivalent to fitting the SFs corresponding to each policy in the policy representation space. Since PeSFA needs to fit the SFs $\tilde{\psi}(s, a, \chi_\pi)$ corresponding to each fixed policy π, it is more reasonable to use the on-policy algorithm.

Therefore, the transition we need to collect is $(s, a, \phi, s', a', \omega_\pi)$, where a' is sampled using the same policy as the selection of action a, ω_π represents the state-action pairs collected by policy π. Training with transitions of this form, we can fit the SFs corresponding to the policy representation on the task w. And the loss function of PeSFA based on TD-error [18] is as follows

$$\mathcal{L}(\tilde{\psi}) = \mathbb{E}\Big[\big(\phi(s,a,s') + \tilde{\psi}_-(s',a',\chi_\pi) - \tilde{\psi}(s,a,\chi_\pi)\big)^2\Big] \tag{8}$$

where $\chi_\pi = \tilde{\Omega}(\omega_\pi)$ is the representation of policy π, $\tilde{\psi}_-$ is the target net of PeSFA.

The process of training PeSFA, and how the gradient ascent based representation search method is used for finding a better policy as well as exploring the representation space more efficiently is provided in Appendix A.

4.2   Theoretical Analysis on Generalizing SFs Among Policies

In this section, we will analyze the generalizability of PeSFA at the policy level and illustrate the difference between using a value-based approach and actor-critic architecture for generalizing the SFs among policy. Then how PeSFA can expedite the policy evaluation and optimization process with the help of policy generalizability will be analyzed.

To better demonstrate our analytical procedure, the approximation loss of PeSFA $\tilde{\psi}_\theta$ at any policy $\pi \in \Pi$ can be expressed as

$$f_\theta(\pi) = \sum_i \|\tilde{\psi}_\theta^i(\pi) - \psi^i(\pi)\| \geq 0, i \in [0,d) \tag{9}$$

Subsequently, similar to the definition of π-Value Approximation in PeVFA [19], the approximation process of PeSFA can be defined as $\mathscr{P}_\pi : \Theta \to \Theta$. and the parameter changes from θ to θ' in this process. Then PeSFA $\tilde{\psi}(\pi)$ will have a more accurate estimate of the SFs $\psi(\pi)$ corresponding to the policy π, and its corresponding loss on the policy π can be a γ-contraction mapping, which can be expressed in the following form

$$f_{\theta'}(\pi) \leq \gamma f_\theta(\pi), \gamma \in [0,1), \theta' = \mathscr{P}_\pi(\theta) \tag{10}$$

We first illustrate what is unique about using value-based methods for policy-level generalization on SFs. For a common action-value function, the policy π is represented implicitly by the value corresponding to the state. When the action-value function is updated, the policy it implies changes. This phenomenon persists when we use the policy representation as an explicit input.

Suppose we train PeSFA on the task w, then the action-value function for the task w is $Q_\theta(s,a,\pi) = \tilde{\psi}_\theta(s,a,\pi)^\top \mathbf{w}$, and subsequently we perform an approximation process on PeSFA of the following form

$$\theta \xrightarrow{\mathscr{P}_\pi} \theta' \tag{11}$$

When performing an approximation process on PeSFA, it can be known from Eq. 10 that $f_{\theta'}(\pi) \leq \gamma f_\theta(\pi)$, i.e., PeSFA fits the SFs better corresponding to the same policy representation. At this point, the action-value function obtained by PeSFA under the parameter θ' is $Q_{\theta'}(s, a, \pi) = \tilde{\psi}_{\theta'}(s, a, \pi)^{\top}w$. Although the same policy representation is input, the policy it implicitly expresses has changed, and this change can be expressed as

$$\exists(s, a), \text{s.t. } \tilde{\psi}_{\theta}(s, a, \pi)^{\top}w \neq \tilde{\psi}_{\theta'}(s, a, \pi)^{\top}w \qquad (12)$$

Therefore, when we perform an approximation of PeSFA on the task w, the policy implied by the action-value function corresponding to the task w changes, i.e., from π to π'. Then the policy representation input to PeSFA should be updated from π to π', which can be computed by $\tilde{\psi}_{\theta'}(s, a, \pi)^{\top}w$. We can then continue the approximation process on the basis of $\tilde{\psi}_{\theta'}(s, a, \pi')^{\top}w$.

Thus the difference between using a value-based approach and actor-critic architecture for generalizing SFs among policy is clear. When the approximation process is applied to SFs, the updated PeSFA is used to calculate the representation of the policy as the new input to SFs, and this step represents the process of policy optimization. If we explicitly train a policy net, the process of policy optimization is to update the policy directly with the value function.

Similar to the above process of approximation and optimization for PeSFA, we define the process of updating PeSFA from θ_{-1} to θ_0 as $\theta_{-1} \overset{\mathscr{P}_{\pi_0}}{\to} \theta_0$. Then, the process of continuously optimizing PeSFA can be expressed as

$$\theta_{-1} \overset{\mathscr{P}_{\pi_0}}{\to} \theta_0 \overset{\mathscr{P}_{\pi_1}}{\to} \theta_1 \overset{\mathscr{P}_{\pi_2}}{\to} \dots \qquad (13)$$

Therefore, we can express the process of optimizing PeSFA consistent with the policy evaluation and optimization process of PeVFA. So the following equation can be obtained according to the theorem of PeVFA

$$f_{\theta_t}(\pi_t) + f_{\theta_t}(\pi_{t+1}) \leq \sum_i \|\psi(s, a, \pi_t) - \psi(s, a, \pi_{t+1})\|, \forall t \geq 0, i \in [0, d) \qquad (14)$$

We can then learn that

$$f_{\theta_t}(\pi_{t+1}) \leq \sum_i \|\tilde{\psi}_{\theta_t}(s, a, \pi_t) - \psi^{\pi_{t+1}}(s, a)\|, i \in [0, d) \qquad (15)$$

According to Eq. 15, the error of predicting the SFs corresponding to π_{t+1} using PeSFA can be smaller than the difference between $\tilde{\psi}_{\theta_t}(s, a, \pi_t)$ and the true SFs $\psi^{\pi_{t+1}}$. Therefore, the generalization of PeSFA allows us to provide a better starting point in each iterative learning. It allows PeSFA to fit the SFs corresponding to each policy representation more quickly.

We will then show that the policy representation points traversed by PeSFA during optimization can form a path, and PeSFA can generalize well among policies over the range of this path. As the process in Eq. 11, the action-value function changes from $\tilde{\psi}_{\theta}(s, a, \chi_{\pi})w$ to $\tilde{\psi}_{\theta'}(s, a, \chi_{\pi})w$, and the policy representation corresponding to the policy also changes from χ_{π} to $\chi_{\pi'}$. Assuming that $f_{\theta}(\pi)$ is L-continuous at policy π, that is

$$|f_{\theta}(\pi) - f_{\theta}(\pi')| \leq L \cdot d(\pi, \pi') \qquad (16)$$

where $\pi' \in \Pi$, d denotes some distance metric in the policy representation space. Since any $t \geq 0$ in Eq. 13, f_{θ_t} is L-continuous in π_t [19]. When starting from a random policy and continuously optimizing PeSFA, we will continue to get better policies through optimizing policy. And finally, the policy obtained in the optimization process can form a path in the policy representation space, then PeSFA generalizes well around this optimization path in the policy representation space.

4.3 Search for Better Policy in the Policy Representation Space

We first briefly introduce the policy representation module. Then we describe how the gradient ascent based representation space search module helps us find a better policy for the corresponding task. In addition, it can also accelerate the learning rate and explore the representation space more efficiently.

We use the surface policy representation (SPR) [19] algorithm with a self-supervised training method in PeVFA as the policy representation module, which expresses the meaning of the policy more refined.

The SPR algorithm takes the state-action pairs $\{(s_i, a_i)\}_{i=1}^{k}$ collected by the policy π as input, and obtain a d-dimensional representation χ_π, this process can be expressed as $\Omega_e(\pi) = \chi_\pi$. Besides, an auxiliary decoder is built to decode the action a corresponding to the policy based on the policy representation χ_π and the state s, i.e., $\Omega_d(s, \chi_\pi) = a$. Then the loss function of the encoder and auxiliary decoder for the policy representation module are as follows

$$\mathcal{L}_{AUX}(\tilde{\Omega}_e, \tilde{\Omega}_d) = \mathbb{E}\left[\left(\tilde{\Omega}_d(\tilde{\Omega}_e(\omega_\pi), a) - a\right)^2\right] \tag{17}$$

where ω_π represents the state-action pairs $\{(s_i, a_i)\}_{i=1}^{k}$ constructed according to the policy π.

For the convenience of training policy representation module, we decide to perform a softmax operation on action-value function $\tilde{\psi}(s, a, \chi_\pi)\mathrm{w}$, which can convert the action into the probability distribution. Then we can construct the state-action pairs as policy representation encoding data in the following form

$$\omega_\pi = \{(s_i, a_i)\}_{i=1}^{k} = \left\{\left(s_i, \frac{\exp(\tilde{\psi}(s_i, \chi_\pi, a)\mathrm{w}/T)}{\sum_j \exp(\tilde{\psi}(s_i, \chi_\pi, a_j)\mathrm{w}/T)}\right)\right\}_{i=1}^{k} \tag{18}$$

With the help of the policy representation module, we can obtain the representation of the policy in a more compact form. On this basis, combining with the search module allows the SFs module to better explore and fit in the representation space and efficiently get a better policy.

Through the training of the algorithm, $\tilde{\psi}(s, a, \chi_\pi)$ fit the corresponding SFs well in the policy representation space Π, that is PeSFA can fit the corresponding SFs plane in the policy representation space very well. Then we can find the SFs $\tilde{\psi}(s, a, \chi_{\pi'_\mathrm{w}})$ of a better policy π'_w for any new task w by searching in the policy representation space corresponding to the SFs plane.

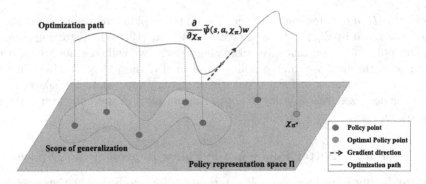

Fig. 2. Optimizing the policy representation in the gradient direction that maximizes the action-value function. (Color figure online)

However, the policy representation space is continuous and infinite. If every policy representation in the space is searched, it will undoubtedly consume a lot of time and not be worth the gain. Therefore, we propose a method that can search more efficiently for SFs of better policy in the policy representation space.

If we have trained on a series of tasks and obtained $\Pi_{history}$, we can choose a historical optimal policy representation point $\chi_\pi \in \Pi_{history}$ or any random policy representation as the starting point χ_0. We can take the gradient direction that maximizes the action-value function $\tilde{\psi}(s, a, \chi_0)w$ as the search direction. Then the policy representation to be optimized is advanced a certain distance in this direction. In addition, a weight β can be used to affect the value of the corresponding action that we expect to maximize, that is $\beta = \frac{\exp(\tilde{\psi}(s,\chi_\pi,a)w/T)}{\sum_j \exp(\tilde{\psi}(s,\chi_\pi,a_j)w/T)}$. To sum up, the method of adjusting the policy representation by gradient direction can be expressed as

$$\chi_{opt} = \chi_\pi + \alpha \frac{\partial}{\partial \chi} \beta \tilde{\psi}(s, a, \chi_\pi)^\top w \qquad (19)$$

Then we can recalculate the gradient direction with the optimized policy representation χ_{opt} as the starting point and move forward a certain distance in this gradient direction. Constantly searching for a better policy representation in the gradient direction can help us find a near-optimal policy representation corresponding to the new task as quickly as possible.

The optimization process of PeSFA and the process of searching for a better policy representation are shown in Fig. 2. The blue circles on the gray plane indicate each policy representation point in the optimization process as in Eq. 13. The black straight line indicates the gradient direction as in Eq. 19. Advancing the policy representation a certain distance in this direction makes the policy representation point that being optimized closer to the optimal policy representation point. Then the figure shows that after the gradient ascent, only a few optimization steps are needed to reach the optimal policy representation point.

At the same time, in the process of training, we can also optimize the representation of the policy that is currently being learned by moving it in the gradient direction. It helps us speed up the learning rate and adjust the learning direction of the current policy to avoid falling into a sub-optimal solution. Besides, the process of optimizing the policy representation can also bring PeSFA from the current range of policy representation space to an unseen policy representation point and continue training. Thereby, the representation searching method can also enhance the generalization ability of the model.

5   Experimental Results

In this section, we present the experimental results of PeSFA and two baselines in the Grid World [3] and Reacher [3,8] environments. First, we will compare the experimental results of PeSFA with two baselines, SFs [3] and USFA [4], in a Grid World environment with discrete state space and the Reacher environment with continuous state space. Then, we will test whether PeSFA combined with the gradient ascent based representation space searching method will affect the experimental results in the Grid world environment.

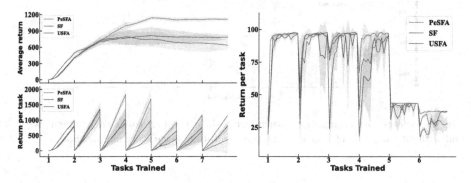

Fig. 3. Left: Average and cumulative return per task in the Four Room environment. **Right:** Average return per task in the Reacher environment.

The first environment is a navigation task in Grid World consisting of four rooms. In this environment, agent starts from a location in a room and needs to reach the goal in another room, where the agent can pick up objects and obtain their corresponding reward by passing through it, similarly as done in [3,8].The second is a continuous state space environment which is constructed on the PyBullet physics engine [5], and the agent can control the robotic arm to reach the preferred target location, similarly as done in [3,8]. We test the algorithm performance on two environments with specific task sets, and more details of the experiments and the hyperparameters are shown in Appendix B.

We can see that PeSFA performs much better than SF and USFA after the second task in the leftmost panel in Fig. 3. And PeSFA has better stability than other algorithms and can achieve better results faster on new tasks. Since PeSFA needs to train the policy representation module at the beginning of the task, the policy encoding part is not trained at the initial time, which leads to poor results for PeSFA at the first task. The rightmost panel in Fig. 3 shows a similar result in the Reacher environment. We can also find that PeSFA has a faster learning rate from the second task onward and can obtain better results at the beginning of the task with the help of finding better policy representation as the initial point.

The experimental results of testing whether PeSFA combines the gradient ascent based representation space searching method in the Grid World environment are shown in Fig. 4. The figure shows that the PeSFA algorithm that combines the searching method has better results on most tasks than the one that does not. We can find that the PeSFA with the search method can get more returns on the task and has a minor variance to get more stable returns. It also shows that by combining the searching method, the PeSFA algorithm can obtain better policy representations for the current task by searching the representation space, thus expediting the learning process.

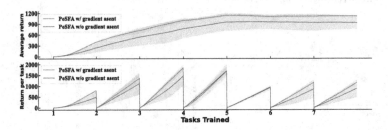

Fig. 4. Average and cumulative return per task in the four-room environment with and without the gradient ascent based representation space searching method.

6   Conclusion

In this paper, we propose to decouple the SFs from the policy by taking the policy representation as an additional input to the SFs. It allows PeSFA to fit the SFs in the policy representation space. At the same time, we devise a method to search for better policy in the policy representation space and exploit the generalizability of PeSFA among task for transferring knowledge of policy. This method helps with continuing training with a better starting point and exploring the policy representation space more efficiently, which leads to a faster learning process and a more efficient knowledge transfer. It is worthwhile investigating whether we can build a PeSFA architecture that can fit SFs under multiple tasks simultaneously to explore the policy representation space more efficiently.

Appendix

A   Training PeSFA

In this section, we will show how to train $\tilde{\psi}(s, a, \chi_\pi)$ in an on-policy way based on the sarsa method.

Algorithm 1 shows the overall process of training PeSFA. First, at the beginning of a task, we will reset the environment and select an action for interacting with the environment (line 4–6), and the action will be executed with some information obtained from the environment, then the next action we will execute will be selected before update the policy (line 9–12). After updating PeSFA, for the reasons described in Sect. 4.1, it is necessary to recalculate the state-action pairs $\omega_{\pi'}$ corresponding to the new policy according to Eq. 18 (line 15–16). We will also search for a better policy in the policy representation space according to Eq. 19, and the χ_{opt} is chosen as the initial policy for subsequent training, which leads to better sample efficiency and exploration in the policy representation space. Besides, we will select the optimized policy representation which is found in the representation space as described in Sect. 4.3.

Algorithm 1. PeSFA

Input: PeSFA net $\tilde{\psi}$, policy representation encoder net $\tilde{\Omega}$, replay buffer D_{memory}

1: **for** $task\_id \leftarrow 1, 2, ..., task\_num$ **do**
2:     initialize χ_π
3:     **for** $epoch \leftarrow 1, 2, ..., max\_epoch$ **do**
4:         select initial state $s \in \mathcal{S}$
5:         **if** $Bernoulli(\epsilon) = 1$ **then** $a \sim Uniform(\{1, 2, ..., |A|\})$
6:         **else** $a \leftarrow \arg\max_b \tilde{\psi}(s, \tilde{\Omega}(\omega_\pi), b)\mathbf{w}$
7:         **end if**
8:         **for** $step \leftarrow 1, 2, ..., max\_step$ **do**
9:             Execute action a and observe s', r, ϕ
10:            **if** $Bernoulli(\epsilon) = 1$ **then** $a' \sim Uniform(\{1, 2, ..., |A|\})$
11:            **else** $a' \leftarrow \arg\max_b \tilde{\psi}(s', \tilde{\Omega}(\omega_\pi), b)\mathbf{w}$
12:            **end if**
13:            push $(s, a, \phi, s', a', \omega_\pi)$ into D_{memory}
14:            $a \leftarrow a'$
15:            update $\tilde{\psi}(s, a, \chi_\pi)$ ▷ see Eq. (8)
16:            sample $states$ from D_{memory}, calculate $\omega_\pi \leftarrow \omega_{\pi'}$ ▷ see Eq. (18)
17:            **if** need to update χ_π by gradient **then**
18:                $\chi_{\pi_{opt}} \leftarrow \chi_\pi + \alpha \frac{\partial}{\partial \chi} \beta \tilde{\psi}(s, a, \chi_\pi)^\top \mathbf{w}$
19:                sample $states$ from D_{memory}, calculate $\omega_\pi \leftarrow \omega_{\pi_{opt}}$ ▷ see Eq. (18)
20:            **end if**
21:         **end for**
22:         update $\tilde{\Omega}$ ▷ see Eq. (17)
23:     **end for**
24: **end for**

B   Additional Experimental Details

For code-level details, our codes are implemented with Python 3.6.9 and Torch 1.11.0. All experiments were run on a single NVIDIA GeForce GTX 1660Ti GPU. The hyperparameters used in Grid World and Reacher experiments are shown in Table 1, and the task weight is shown in Table 2.

The first experimental environment is a navigation task in Grid World, a two-dimensional discrete space consisting of four rooms. In this environment, the agent starts from a location in a room and needs to reach a goal point in another room, where the agent can pick up objects and obtain their corresponding reward by passing through it, similarly as done in [3,8]. These objects belong to one of the three types of objects and each type of object has a specific reward. The location of each object in the environment remains the same for all tasks, but the reward of each type of object varies with the task. The goal is to maximize the cumulative sum of reward values over tasks. And ϕ and w are artificially constructed, which satisfy the reward function in Eq. 2, and $\phi \in \mathbb{R}^4$ represents whether a particular type of object is passed in that transition and $w \in \mathbb{R}^4$ indicates the reward corresponding to each type of objects.

The second environment is a continuous state space environment constructed on the PyBullet physics engine [5], and the agent can control the robotic arm to reach the preferred target location, as done in [3,8]. In each task of this environment, we control the degree of preference to each target locus by controlling the task weights w, while $\phi \in \mathbb{R}^4$ represents the negative of the euclidean distance from the robotic arm's tip to each target locus and then adding one.

Table 1. PeSFA's hyperparameters per environment.

	Grid world	Reacher
PeSFA network, $\tilde{\psi}$	MLP([256, 256])	MLP([512, 512])
Minibatch size	64	256
Learning rate	0.001	0.001
Gamma	0.95	0.9
Optimiser	ADAM	ADAM
Policy representation dim	12	6
α	0.0003	0.0003
SPR encoder network $\tilde{\Omega}_e$	MLP([64, 64])	MLP([64, 64])
SPR decoder network $\tilde{\omega}_d$	MLP([256, 256])	MLP([256, 256])

Table 2. Task weight per environment.

Task	Grid world	Reacher
1	$[0.25, 0.25, 0.25]$	$[1, 0, 0, 0]$
2	$[1., 0., 0.]$	$[0, 1, 0, 0]$
3	$[0., 1., 0.]$	$[0, 0, 1, 0]$
4	$[0., 0., 1.]$	$[0, 0, 0, 1]$
5	$[-0.25, 0.25, 0.25]$	$[0.25, 0.25, 0.25, 0.25]$
6	$[0.25, -0.25, 0.25]$	$[-0.25, 0.25, 0.25, 0.25]$
7	$[0.25, 0.25, -0.25]$	

References

1. Alegre, L.N., Bazzan, A.L.C., da Silva, B.C.: Optimistic linear support and successor features as a basis for optimal policy transfer. In: Chaudhuri, K., Jegelka, S., Song, L., Szepesvári, C., Niu, G., Sabato, S. (eds.) International Conference on Machine Learning, ICML 2022, Baltimore, Maryland, USA, 17–23 July 2022. Proceedings of Machine Learning Research, vol. 162, pp. 394–413. PMLR (2022)
2. Alver, S., Precup, D.: Constructing a good behavior basis for transfer using generalized policy updates. In: The Tenth International Conference on Learning Representations, ICLR 2022, Virtual Event, 25–29 April 2022. OpenReview.net (2022)
3. Barreto, A., et al.: Successor features for transfer in reinforcement learning. In: Guyon, I., et al. (eds.) Advances in Neural Information Processing Systems 30: Annual Conference on Neural Information Processing Systems 2017, Long Beach, CA, USA, 4–9 December 2017, pp. 4055–4065 (2017)
4. Borsa, D., et al.: Universal successor features approximators. CoRR abs/1812.07626 (2018)
5. Ellenberger, B.: Pybullet gymperium (2018–2019)
6. Feinberg, A.: Markov decision processes: discrete stochastic dynamic programming (Martin l. Puterman). SIAM Rev. 38(4), 689 (1996)
7. Filos, A., Lyle, C., Gal, Y., Levine, S., Jaques, N., Farquhar, G.: Psiphi-learning: reinforcement learning with demonstrations using successor features and inverse temporal difference learning. In: Meila, M., Zhang, T. (eds.) Proceedings of the 38th International Conference on Machine Learning, ICML 2021, 18–24 July 2021, Virtual Event. Proceedings of Machine Learning Research, vol. 139, pp. 3305–3317. PMLR (2021)
8. Gimelfarb, M., Barreto, A., Sanner, S., Lee, C.: Risk-aware transfer in reinforcement learning using successor features. In: Ranzato, M., Beygelzimer, A., Dauphin, Y.N., Liang, P., Vaughan, J.W. (eds.) Advances in Neural Information Processing Systems 34: Annual Conference on Neural Information Processing Systems 2021, NeurIPS 2021, 6–14 December 2021, Virtual, pp. 17298–17310 (2021)
9. Han, D., Tschiatschek, S.: Option transfer and SMDP abstraction with successor features. In: Raedt, L.D. (ed.) Proceedings of the Thirty-First International Joint Conference on Artificial Intelligence, IJCAI 2022, Vienna, Austria, 23–29 July 2022, pp. 3036–3042. ijcai.org (2022)
10. Hansen, S., Dabney, W., Barreto, A., Warde-Farley, D., de Wiele, T.V., Mnih, V.: Fast task inference with variational intrinsic successor features. In: 8th Inter-

national Conference on Learning Representations, ICLR 2020, Addis Ababa, Ethiopia, 26–30 April 2020. OpenReview.net (2020)

11. Lillicrap, T.P., et al.: Continuous control with deep reinforcement learning. In: Bengio, Y., LeCun, Y. (eds.) 4th International Conference on Learning Representations, ICLR 2016, San Juan, Puerto Rico, 2–4 May 2016, Conference Track Proceedings (2016)

12. Liu, H., Abbeel, P.: APS: active pretraining with successor features. In: Meila, M., Zhang, T. (eds.) Proceedings of the 38th International Conference on Machine Learning, ICML 2021, 18–24 July 2021, Virtual Event. Proceedings of Machine Learning Research, vol. 139, pp. 6736–6747. PMLR (2021)

13. Mnih, V., et al.: Human-level control through deep reinforcement learning. Nat. **518**(7540), 529–533 (2015)

14. Nemecek, M.W., Parr, R.: Policy caches with successor features. In: Meila, M., Zhang, T. (eds.) Proceedings of the 38th International Conference on Machine Learning, ICML 2021, 18–24 July 2021, Virtual Event. Proceedings of Machine Learning Research, vol. 139, pp. 8025–8033. PMLR (2021)

15. Raileanu, R., Goldstein, M., Szlam, A., Fergus, R.: Fast adaptation to new environments via policy-dynamics value functions. In: Proceedings of the 37th International Conference on Machine Learning, ICML 2020, 13–18 July 2020, Virtual Event. Proceedings of Machine Learning Research, vol. 119, pp. 7920–7931. PMLR (2020)

16. Schaul, T., Horgan, D., Gregor, K., Silver, D.: Universal value function approximators. In: Bach, F.R., Blei, D.M. (eds.) Proceedings of the 32nd International Conference on Machine Learning, ICML 2015, Lille, France, 6–11 July 2015. JMLR Workshop and Conference Proceedings, vol. 37, pp. 1312–1320. JMLR.org (2015)

17. Silver, D., et al.: Mastering the game of go with deep neural networks and tree search. Nat. **529**(7587), 484–489 (2016)

18. Sutton, R.S., Barto, A.G.: Reinforcement Learning - An Introduction. Adaptive Computation and Machine Learning. MIT Press (1998)

19. Tang, H., et al.: What about inputing policy in value function: policy representation and policy-extended value function approximator. In: Thirty-Sixth AAAI Conference on Artificial Intelligence, AAAI 2022, Thirty-Fourth Conference on Innovative Applications of Artificial Intelligence, IAAI 2022, The Twelveth Symposium on Educational Advances in Artificial Intelligence, EAAI 2022 Virtual Event, 22 February–1 March 2022, pp. 8441–8449. AAAI Press (2022)

20. Taylor, M.E., Stone, P.: Transfer learning for reinforcement learning domains: a survey. J. Mach. Learn. Res. **10**, 1633–1685 (2009). https://dl.acm.org/doi/10.5555/1577069.1755839

21. Yang, T., et al.: Efficient deep reinforcement learning via adaptive policy transfer. In: Proceedings of the Twenty-Ninth International Joint Conference on Artificial Intelligence, pp. 3094–3100 (2020)

22. Zhu, Z., Lin, K., Zhou, J.: Transfer learning in deep reinforcement learning: a survey. CoRR abs/2009.07888 (2020)

Maximal Information Propagation
with Limited Resources

Xu Ge, Xiuzhen Zhang, and Dengji Zhao[✉]

ShanghaiTech University, Shanghai, China
{gexu,zhangxzh1,zhaodj}@shanghaitech.edu.cn

Abstract. We consider an information propagation game, where the sponsor holds the information and wants to attract more players with a fixed resource. We propose an allocation mechanism to incentivize the existing players to propagate the information to more friends. The incentives come from the fact that a player will share more when she propagates more, but her share is also reduced by the others' propagation. Under the new allocation mechanism, for each player, propagating to all her friends is a dominant strategy. The mechanism offers a new perspective for advertising with a limited budget and has great potential in practice as people can easily reach each other via their social platforms.

Keywords: Information propagation · Mechanism design · Social network

1 Introduction

Social networks, formed by agents' connections, are popular platforms for rapid information propagation and exchanges. Utilizing this characteristic, mechanism design in social networks [20,21] finds applications in auctions [10,11,22], online marketing [7,8], answer querying [4,9,18], blockchains [6,12], cooperative games [19] and so on. These can all be modelled as the information propagation game where the goal of the mechanism designer is to incentivize agents to diffuse some particular information to their neighbours, thus making as many agents as possible be informed. However, there is competition among agents, so the participation of new agents may bring loss to existing agents. Therefore, the challenge here is that strategic agents are not willing to diffuse information to others unless they can gain from the diffusion. Thus, one straightforward method is to reward agents for their diffusion.

In the information propagation game, agents' rewards usually come from the divisible resource provided by the sponsor. Then, the challenge remains since the limited resource is shared among more agents with information propagation. There have been many studies working on this. Most of them focused on the scenarios where the sponsor's goal for information propagation is to perform some tasks, e.g., finding an answer [3,9,16] or making a bitcoin transaction [1]. In

M. Yokoo et al. (Eds.): DAI 2022, LNAI 13824, pp. 45–59, 2023.
https://doi.org/10.1007/978-3-031-25549-6_4

these settings, agents' rewards are determined by their contributions to performing the task, and agents with non-zero rewards actually form a single path. In the scenario where the sponsor just wants to maximize the information propagation, agents diffusing information should all have chances to be rewarded. Then, the aforementioned works cannot take full advantage of the provided resource. One work was proposed to incentivize information propagation with a limited monetary budget [14]. However, when the resource is heterogeneous, i.e., agents' valuation functions on the resources can differ, it becomes a cake-cutting problem [2,17]. Another challenge immediately appearing is that strategic agents may misreport their valuation functions to get more [13].

Therefore, in this paper, we want to design a cake-cutting mechanism that provides agents both incentives to propagate the information to all their neighbours and truthfully report their valuation functions. Though these two incentives don't conflict with each other, things don't get easier even when we only guarantee one of them. When we only consider the propagation incentive, the difficulty is that the cake is shared among all participants, and we cannot offer the invitation incentives with extra monetary rewards. When we only require the incentive to report valuation functions truthfully, it becomes a traditional cake-cutting problem. In traditional cake-cutting settings, proportionality is one important fairness criterion, requiring that each agent receives a part that she values at least $1/n$ if the cake is divided among n agents [15]. However, to our best knowledge, no deterministic, proportional and incentive compatible cake-cutting mechanism is proposed without constraints for agents' valuation functions [2,5,13]. We also prove that proportionality cannot be satisfied if we require the propagation incentive since agents' share is decreasing when n is increasing.

To achieve the above objectives, we propose a cake-cutting mechanism to incentivize the information propagation and report valuation functions truthfully, called the *Invitation Incentive Cake-cutting Mechanism*. To combat the issue that agents are unwilling to diffuse information, we divide all participants into layers. Then, we introduce the priority between layers and competitions among agents in the same layer. More precisely, an agent closer to the sponsor has a higher priority to share the resource. For agents in the same layer, an agent gets more from the other agents if she diffuses more. Regarding the issue that agents may strategically report their valuation functions, we introduce randomization when deciding the allocation. Though proportionality cannot be achieved in our setting, we propose a local fairness to give a minimum guarantee for each agent's expected utility, which also guarantees the propagation incentive.

The remainder of the paper is organized as follows. We begin by introducing our basic settings and definitions in Sect. 2. Our mechanism and an example are presented in Sect. 3. We show the properties of our mechanism in Sect. 4. We conclude our work in Sect. 5.

2   The Model

There is a group of n agents $N = \{1, ..., n\}$ and a sponsor s in the network. The sponsor s owns a divisible resource, which can be represented by a cake, denoted by an interval $\mathcal{C} = [0,1]$. Each agent $i \in N$ has a private integrable valuation density function $f_i : [0,1] \rightarrow \mathbb{R}_{\geq 0}$ over the cake. A share of cake $S \subseteq \mathcal{C}$ is denoted by union of several sub-intervals. For a share S, agent i's valuation is $v_i(S) = \int_{x \in S} f_i(x)\mathrm{d}x$. In this way, the valuation is additive, which means that for any two non-overlapping shares S_1 and S_2, there is $v_i(S_1 \cup S_2) = v_i(S_1) + v_i(S_2)$. To simplify calculations, we normalize agents' valuation for the entire cake as $v_i(\mathcal{C}) = 1$ for each agent $i \in N$.

Given the above settings, the sponsor's goal is to propagate the information to as many agents as possible, with a limited divisible resource. The sponsor only connects to some of the agents, and only these agents can be informed initially. These agents also connect to some of the other agents so that they could propagate the information to their neighbours. However, inviting others to share together is not beneficial for them because this may reduce their allocated share. Furthermore, agents may misreport their valuation functions to get more rewards, which may affect the actual propagation results for the sponsor. Therefore, in this paper, our goal is to design a cake-cutting mechanism where agents are incentivized to propagate the information to all their neighbours and truthfully report their valuation functions.

We give formal definition of the relationship among agents. The relationship forms a directed graph $G = (V, E)$, where the sponsor s is the only source. In the graph, the node set is $V = \{s\} \cup N$. Each edge $(i, j) \in E$ indicates that agent j is a neighbour of i, so i can directly propagate to j. For each node $i \in V$, the set of her neighbours is $r_i = \{j \in N \mid (i,j) \in E\}$. For each agent $i \in N$, her type is denoted by $t_i = (r_i, f_i)$, containing the set of her neighbours and her valuation to the cake. The type space of each agent is $\mathcal{P}(N) \times \mathcal{F}$, where $\mathcal{P}(N)$ is the power set of N, and $\mathcal{F}$ is the set of all valuation density functions.

Let $\mathbf{t} = (t_i)_{i \in N}$ be the type profile of all agents, $\mathcal{T}$ be the space of all agents' type profiles. Let $t'_i = (r'_i, f'_i)$ be the type report of agent i. Note that agent i cannot know other agents who are not her neighbours in the network, so we have $r'_i \subseteq r_i$. The space of all possible reports of agent i is denoted by $\mathcal{P}(r_i) \times \mathcal{F}$. The report profile of all agents is denoted as $\mathbf{t}' = (t'_i)_{i \in N}$. $\mathbf{t}'$ is also represented by $(t'_i, \mathbf{t}'_{-i})$, where $\mathbf{t}'_{-i}$ is the report profile of all other agents except for i. In the following, we give some important definitions of the cake-cutting mechanism.

Definition 1. *Given a report profile* $\mathbf{t}'$ *of all agents, define the* **propagation network** *as* $G(\mathbf{t}') = (V(\mathbf{t}'), E(\mathbf{t}'))$, *where* $V(\mathbf{t}') = \{i_m \mid$ *there exists a sequence:* $s, i_1, i_2, \cdots, i_m$, *where* $i_1 \in r_s$ *and* $i_{k+1} \in r'_{i_k}$ *for all* $1 \leq k < m\}$ *and* $E(\mathbf{t}') = \{(i,j) \mid i \in V(\mathbf{t}'), j \in r'_i\} \cup \{(s,j) \mid j \in r_s\}$.

In the propagation network, agent i can only be informed when there is at least one information propagation path from the sponsor s to agent i. Therefore, with all agents' report profiles, we can find all agents who are informed of the information. All the other agents are not included in the propagation network, but their reports can be considered as potential reports if they are invited.

Definition 2. *Given an propagation network constructed from a report profile* $\mathbf{t}'$, *let the depth of an agent* i, *denoted as* d_i, *be the length of the shortest path* from s to i. The k-th layer l_k is defined as the set of all agents with depth k, i.e., $l_k = \{i \in V(\mathbf{t}') \mid d_i = k\}$.

Definition 3. *Given a report profile* $\mathbf{t}'$ *and* $G(\mathbf{t}') = (V(\mathbf{t}'), E(\mathbf{t}'))$, *for agent* $i \in l_k$, *let* $\tilde{r}_i(\mathbf{t}')$ *be the set of edges from* i *to some agent in the layer* l_{k+1} *when agent* i *report* $\mathbf{t}'$, *i.e.,* $\tilde{r}_i(\mathbf{t}') = \{(i,j) \mid i \in l_k, j \in l_{k+1}, (i,j) \in E(\mathbf{t}')\}$. *Let* $R_k(\mathbf{t}') = \bigcup_{i \in l_k} \tilde{r}_i(\mathbf{t}')$ *be the set of all edges from the* k-th *layer to the* $(k+1)$-th *layer. When* $\mathbf{t}'$ *is clear in the context, we simply write* $\tilde{r}_i$ *and* R_k.

We provide an example to illustrate the definition of layers and edge sets. According to the report profile $\mathbf{t}'$, we construct a directed graph as the propagation network. Based on agents' depths in the propagation network, we divide the agents into different layers and find the edges between adjacent layers. For example, in Fig. 1, the max depth of the graph is 3 and agents are divided into three layers l_1, l_2 and l_3. Agent 3, agent 4 and agent 5 have the same depth 2 so that they are in the same layer. For agent 5, the set of edges from her to agents in the next layer is $\tilde{r}_5 = \{(5,7)\}$. Then, the set of edges from the layer l_2 to the layer l_3 is $R_2 = \{(3,6),(4,6),(4,7),(5,7)\}$.

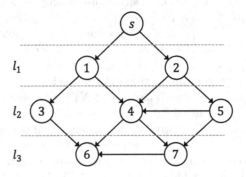

Fig. 1. An example for illustrating the definition of the network layer and the set of edges, where $l_1 = \{1,2\}, l_2 = \{3,4,5\}, l_3 = \{6,7\}$, and $\tilde{r}_5 = \{(5,7)\}$. The edge set from the layer l_2 to the layer l_3 is denoted by $R_2 = \{(3,6),(4,6),(4,7),(5,7)\}$.

In our problem setting, an allocation can be represented by a vector containing n shares. An allocation should fully allocate the cake to agents, and there must be no overlap among each agent's share. We give the formal definition of an allocation and a cake-cutting mechanism in the following.

Definition 4. *An **allocation** of the cake* $\mathcal{C} = [0,1]$ *among* N *is defined as* $\mathbf{A} = (A_i)_{i \in N}$, *where* $A_i \cap A_j = \emptyset$ *for any* $i, j \in N$ *and* $\bigcup_{i \in N} A_i = \mathcal{C}$.

Definition 5. *A **cake-cutting mechanism** is defined by* $\mathcal{M} : \mathcal{T} \to \mathcal{A}$, *where* $\mathcal{A}$ *is the space of all allocations. Given the set of agents* N *and their report profile* $\mathbf{t}' = (t'_i)_{i \in N}$, *the output of such mechanism is an allocation* $\mathbf{A} = (A_1, \cdots, A_n)$, *where* A_i *is the share allocated to agent* i.

For all possible report profile $\mathbf{t}'$, *a cake-cutting mechanism satisfies that 1) for each agent* $i \notin V(\mathbf{t}')$, $A_i = \emptyset$, *and 2) for each agent* $i \in V(\mathbf{t}')$, A_i *is independent of the reports of the agents that are not in* $V(\mathbf{t}')$.

Next, we define the property of incentive compatibility for a cake-cutting mechanism in our setting, which incentivizes agents to report their types truthfully.

Definition 6. *A cake-cutting mechanism* $\mathcal{M}$ *is* ***incentive compatible*** *if for all* $\mathbf{t} \in \mathcal{T}$, *all* $i \in N$, *all* $t_i' \in \mathcal{P}(r_i) \times \mathcal{F}$ *and all* $\mathbf{t}'_{-i} \in \times_{j \in N \setminus \{i\}} (\mathcal{P}(r_j) \times \mathcal{F})$,

$$\mathbb{E}\left(v_i\left(\mathcal{M}\left((t_i, \mathbf{t}'_{-i})\right)_i\right)\right) \geq \mathbb{E}\left(v_i\left(\mathcal{M}\left((t_i', \mathbf{t}'_{-i})\right)_i\right)\right).$$

The property of incentive compatibility guarantees that for each agent $i \in N$, it is a dominant strategy to invite all her neighbours and report her true valuation function. Below, we provide the definition of proportionality in traditional cake-cutting settings.

Definition 7. *A cake-cutting mechanism* $\mathcal{M}$ *is* ***proportional*** *if* $v_i(A_i) \geq \frac{1}{n}$ *for* $i \in N$.

Then, we show that no cake-cutting mechanism can satisfy incentive compatibility and proportionality simultaneously when considering information propagation.

Proposition 1. *In the network setting, no cake-cutting mechanism can satisfy incentive compatibility and proportionality simultaneously.*

Proof. We prove this conclusion by showing that an agent may misreport her neighbours to increase her utility. Assume that all agents share the same valuation density function for the cake. Then the only proportional distribution is that each agent receives exactly $1/n$ of the entire cake.

Suppose that propagation network $G(\mathbf{t})$ contains at least two layers. Consider the action vector where *each agent in the first layer do not invite any neighbour.* Under this action, each of them will get $\frac{1}{|l_1|}$. If one of the agents changes her action to inviting k neighbours ($k > 0$), then her piece will reduce to $\frac{1}{|l_1|+k}$. In this way, truthful propagation leads to a decreasing utility, so no proportional mechanism can be incentive compatible.

3   The Mechanism

In this section, we present the Invitation Incentive Cake-cutting Mechanism (IICM). The intuition of our mechanism is that we assign different priorities to agents in different layers such that agents will not get smaller shares from diffusion. Before giving the mechanism, we first introduce several notations used in our mechanism.

Given a report profile $\mathbf{t}'$ of all agents and the generated propagation network $G(\mathbf{t}')$, assume that there are total m layers. Denote them as $l_1, l_2, \cdots, l_m$.

- For all agents with their depth no less than k, i.e., $i \in \bigcup_{j=k}^{m} l_j$, denotes the share that divided among them as $\mathcal{C}_k$. Then, $\mathcal{C}_1$ denotes the entire cake.
- For agent i with depth k, let X_i be the initial share assigned to her after dividing $\mathcal{C}_k$ among agents in l_k.
- For an edge $(i,j) \in R_k$, let $Y_i^{(i,j)}$ be the share rewarding i for inviting j, and $Y_j^{(i,j)}$ be the share given to j's layer.

As for input parameters α, β, γ, these are hyperparameters of the algorithm, which are required to be positive integers and satisfy $\alpha \geq \beta + \gamma$. The parameters are used to tune how the reward is shared between adjacent layers and among agents in the same layer. We will illustrate the effect of parameters later in Example 1.

Then, we can give the definition of the Invitation Incentive Cake-cutting Mechanism, as shown in Algorithm 1.

Algorithm 1. Invitation Incentive Cake-cutting Mechanism

Input: A report profile $\mathbf{t}'$, a cake $\mathcal{C}$ and parameters $\alpha, \beta, \gamma \in \mathbb{N}_+$, where $\alpha \geq \beta + \gamma$.
Output: An allocation $\mathbf{A} = (A_1, \cdots, A_n)$ of the cake $\mathcal{C}$.
1: Construct the propagation network $G(\mathbf{t}')$.
2: Divide agents into different layers according to their depth. Let $m = \max_{i \in N} d_i$.
3: Set $A_i \leftarrow \emptyset$ for all $i \in N$, $\mathcal{C}_1 \leftarrow \mathcal{C}$ and $\mathcal{C}_k \leftarrow \emptyset$ for all $1 < k \leq m$.
4: **for** each layer l_k, where $k = 1, 2, ..., m$ **do**
5:   Randomly divide $\mathcal{C}_k$ into $|l_k|$ parts, then assign one part to each agent $i \in l_k$ as her initial share X_i.
6:   Each agent $i \in l_k$ divides her initial share X_i into $\alpha \cdot |R_k|$ parts with equal values for i.
7:   **for** each edge $(i,j) \in R_k$ **do**
8:     **for** each agent $i' \in l_k \setminus \{i\}$ **do**
9:       Randomly choose β parts as $Y_i^{(i,j)}$, choose γ parts as $Y_j^{(i,j)}$, both from $X_{i'}$. Update $A_i, \mathcal{C}_{k+1}, X_{i'}$ as follows.
10:       $A_i \leftarrow A_i \cup Y_i^{(i,j)}$.
11:       $\mathcal{C}_{k+1} \leftarrow \mathcal{C}_{k+1} \cup Y_j^{(i,j)}$.
12:       $X_{i'} \leftarrow X_{i'} \setminus \left(Y_i^{(i,j)} \cup Y_j^{(i,j)} \right)$.
13:     **end for**
14:   **end for**
15:   For each agent $i \in l_k$, $A_i \leftarrow A_i \cup X_i$, i.e., allocate the retained shares to i.
16: **end for**
17: **return** $(A_i)_{i \in N}$ as the allocation of the cake $\mathcal{C}$.

The main idea of the Invitation Incentive Cake-cutting Mechanism is that we introduce the concept of priority and competition in layers when distributing the cake. One obstacle to achieving incentive compatibility in social networks is that agents may get smaller shares if they are with the same priority. Therefore, our mechanism assigns higher priority to the agents that are closer to the sponsor so that they can choose before other agents. In addition, competition should be encouraged among agents in the same layer, as this will incentivize them to invite their neighbours, even if some of the shares will flow to the following layers. In our mechanism, for each agent in layer l_k, they will divide their initial share into $\alpha \cdot |R_k|$ parts with the same value for themselves. For each directed edge $(i,j) \in R_k$, agent i gets a reward of β parts, and γ parts will be assigned to the next layer. The aforementioned $\beta + \gamma$ parts come from the initial share of other agents in i's layer.

Our mechanism provides agents having more propagation contributions with more rewards, and the rewards come from other agents in the same layer. Then, all agents will invite all their neighbours such that they would win a larger share.

3.1   An Example of Our Mechanism

In the following, we give an example to illustrate the running process of the Invitation Incentive Cake-cutting Mechanism. Consider a social network shown in Fig. 2. The sponsor s has two neighbours, agent 1 and agent 2. Agent 1 and agent 2 have a common neighbour, agent 3. Then, we have $r_1 = r_2 = \{3\}$ and $r_3 = \emptyset$. The sponsor has a cake $\mathcal{C}$. The graph of agents' valuation density functions for the cake are also shown in Fig. 2. The valuation density function of agent 1 is as the following.

$$f_1(x) = \begin{cases} 0 & 0.1 \leq x \leq 0.85 \\ 4 & \text{otherwise.} \end{cases}$$

For agent 2 and agent 3, their valuation density functions are both uniform distribution, i.e., $f_2(x) = f_3(x) = 1$.

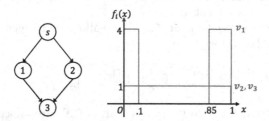

Fig. 2. Left: A social network containing a sponsor and three agents. Right: The valuation density functions of all agents in the social network. Agent 1 has an nonuniform valuation density valuation function, while agent 2 and 3 both have uniform ones.

We set $\alpha = 2$ and $\beta = \gamma = 1$. As for randomly choosing shares, we always choose the leftmost share out of the remaining shares. Here, we first consider the case where agents are all truthfully reporting their types, then consider the cases where they misreport the neighbours.

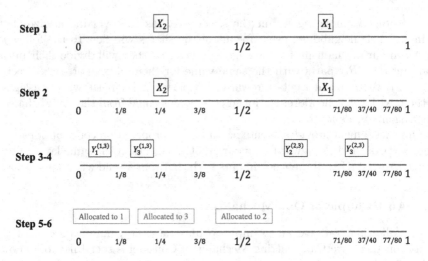

Fig. 3. The running procedure of our mechanism with $\alpha = 2, \beta = \gamma = 1$, where all agents truthfully report. In the final step, the blue parts are allocated to agent 1, the red parts are allocated to agent 2 and the green part are allocated to agent 3. (Color figure online)

The Case When All Agents Truthfully Report. To complete the allocation of all agents, our mechanism runs in six steps as follows, which is shown in Fig. 3.

- **Step 1:** Suppose C is cut at $\frac{1}{2}$, and the initial shares to agent 1 and agent 2 are $X_1 = \left[\frac{1}{2}, 1\right], X_2 = \left[0, \frac{1}{2}\right]$.
- **Step 2:** Agent 1 and 2 are in the layer l_1 and we have $R_1 = \{(1,3), (2,3)\}$. Therefore, they are required to cut the initial shares into $\alpha \cdot |R_1| = 4$ equal parts (i.e., their shares are cut at 3 points in their intervals). Then, X_1 is cut at $\frac{71}{80}, \frac{37}{40}, \frac{77}{80}$, and X_2 is cut at $\frac{1}{8}, \frac{1}{4}, \frac{3}{8}$.
- **Step 3:** For edge $(1,3)$, two parts of X_2 are allocated to agent 1 and the next layer respectively, and we have

$$Y_1^{(1,3)} = \left[0, \frac{1}{8}\right], Y_3^{(1,3)} = \left[\frac{1}{8}, \frac{1}{4}\right].$$

- **Step 4:** Similarly, for edge $(2,3)$, two parts of X_1 are allocated to agent 2 and the next layer respectively, and we have

$$Y_2^{(2,3)} = \left[\frac{1}{2}, \frac{71}{80}\right], Y_3^{(2,3)} = \left[\frac{71}{80}, \frac{37}{40}\right].$$

- **Step 5:** According to the mechanism, agent 3 gets the whole C_2, where

$$C_2 = Y_3^{(1,3)} \cup Y_3^{(2,3)} = \left[\frac{1}{8}, \frac{1}{4}\right] \cup \left[\frac{71}{80}, \frac{37}{40}\right].$$

- **Step 6:** For agents 1 and 2, their allocation is the union of the propagation reward shares and the remaining parts of initial shares. Agent 3 is the unique agent in l_2, so she gets the whole C_2. Therefore, the allocation should be as follows.

$$A_1 = \left(X_1 \backslash \left(Y_2^{(2,3)} \cup Y_3^{(2,3)} \right) \right) \cup Y_1^{(1,3)} = \left[0, \frac{1}{8} \right] \cup \left[\frac{37}{40}, 1 \right],$$

$$A_2 = \left(X_2 \backslash \left(Y_1^{(1,3)} \cup Y_3^{(1,3)} \right) \right) \cup Y_2^{(2,3)} = \left[\frac{1}{4}, \frac{71}{80} \right],$$

$$A_3 = C_2 = \left[\frac{1}{8}, \frac{1}{4} \right] \cup \left[\frac{71}{80}, \frac{37}{40} \right].$$

For each agent, the valuation of her own share is as follows.

$$v_1(A_1) = 0.7, v_2(A_2) = 0.6375, v_3(A_3) = 0.1625.$$

The Cases When Agents Misreport Their Neighbours. In the social network shown in Fig. 2, agent 1 and agent 2 can misreport by not inviting anent 3. We list all the three cases where agents misreport in the following.

1. Agent 1 does not invite agent 3.
2. Agent 2 does not invite agent 3.
3. Both agent 1 and agent 2 do not invite agent 3.

For these cases, we omit the running details of our mechanism here and directly give the allocation results instead. The results are shown below.

- **Case 1:** The allocation is $A_1 = \emptyset$, $A_2 = \left[0, \frac{37}{40} \right]$ and $A_3 = \left[\frac{37}{40}, 1 \right]$. We have

$$v_1(A_1) = 0, v_2(A_2) = 0.925, v_3(A_3) = 0.075$$

- **Case 2:** The allocation is $A_1 = \left[0, \frac{1}{4} \right] \cup \left[\frac{1}{2}, 1 \right]$, $A_2 = \emptyset$, $A_3 = \left[\frac{1}{4}, \frac{1}{2} \right]$. We have

$$v_1(A_1) = 1, v_2(A_2) = 0, v_3(A_3) = 0.25$$

- **Case 3:** The allocation is $A_1 = \left[\frac{1}{2}, 1 \right]$, $A_2 = \left[0, \frac{1}{2} \right]$, $A_3 = \emptyset$. We have

$$v_1(A_1) = 0.6, v_2(A_2) = 0.5, v_3(A_3) = 0$$

Combining the truthful reporting case and the three misreporting cases, there is a game on whether inviting agent 3 between agent 1 and agent 2. The payoff matrix are summarized in Table 1.

As we can see, the result is consistent with the envisioned fact, where inviting agent 3 is a dominant strategy for both agent 1 and agent 2. Therefore, the agents are incentivized to propagate the information in our mechanism.

Table 1. The payoff matrix of agent 1 and agent 2.

Agent 1 \ Agent 2	Invite Agent 3	NOT Invite Agent 3
Invite Agent 3	0.6375 0.7	0 1
NOT Invite Agent 3	0.925 0	0.5 0.6

4 Properties of IICM

In this section, we will prove that the Invitation Incentive Cake-cutting Mechanism satisfies incentive compatibility. Furthermore, the Invitation Incentive Cake-cutting Mechanism can also give each agent a minimum guarantee for her share in her layer. Before discussing the properties, we first give two lemmas.

Lemma 1. *Suppose one share X is cut into m parts, and we randomly choose k parts $(m, k \in \mathbb{N}_+, k \leq m)$. Then, for each agent $i \in N$, the expected valuation of these k parts is $\frac{k}{m} v_i(X)$.*

Proof. Let the m parts of X be $x_1, x_2, \cdots, x_m$. To randomly select k parts, each part is selected with a probability of $\frac{k}{m}$. Then, for agent i, the expected valuation of these k parts is

$$\frac{k}{m} \left(v_i(x_1) + v_i(x_2) + \cdots + v_i(x_m) \right) = \frac{k}{m} \left(v_i(x_1 \cup x_2 \cup \cdots \cup x_m) \right) = \frac{k}{m} \cdot v_i(X)$$

Lemma 1 states that when the shares are randomly selected, the expected valuation for each agent only depends on the number of shares. Then, the following lemma gives the portion of agent i's initial share and other agents except i's initial share in the layer's whole share.

Lemma 2. *Applying the Invitation Incentive Cake-cutting Mechanism, for all $k = 1, 2, \cdots, m$ and all $i \in l_k$, the share C_k is independent of i's report profile t_i'. There are also $\mathbb{E}(v_i(X_i)) = \frac{1}{|l_k|} v_i(C_k)$ and $\mathbb{E}\left(\sum_{j \in l_k \setminus \{i\}} v_i(X_j)\right) = \frac{|l_k|-1}{|l_k|} v_i(C_k)$.*

Proof. We first prove the independence of C_k and $t_i' = (r_i', f_i')$. As for the valuation function f_i', it is just used for deciding how i cuts her initial share later, hence cannot affect C_k. So we should show that first k layers $l_1, l_2, ..., l_k$ is not related to r_i'. Intuitively, this conclusion holds because IICM runs in breadth-first order. However, we will give the formal proof in the following.

Consider two groups of agents: agents in the first k layers, and other agents. Given a propagation network $G(\mathbf{t}')$, for each agent $j \in N$, if j is in the first k layers, the shortest path from s to j cannot include i, so her depth d_j is independent of r_i'. If j is not in the first k layers, j will not enter the first k

layers no matter whether there exist an information diffusion path from i to her. Assume that j enters first k layers because of i's invitation, then the path from s to i to j with length no more than k. It indicates that i is in first $(k-1)$ layers, which is a contradiction.

After $\mathcal{C}_k$ is generated, it is divided into $|l_k|$ parts, then each agent in l_k randomly gets one part. With Lemma 1, we have $\mathbb{E}(v_i(X_i)) = \frac{1}{|l_k|}v_i(\mathcal{C}_k)$. As for other agents' initial shares, it is equivalent to choosing $|l_k| - 1$ parts among them. Using Lemma 1, the expected value is $\mathbb{E}\left(\sum_{j\in l_k\backslash\{i\}} v_i(X_j)\right) = |l_k\backslash\{i\}| \cdot \frac{1}{|l_k|}v_i(\mathcal{C}_k) = \frac{|l_k|-1}{|l_k|}v_i(\mathcal{C}_k)$.

In brief, we have shown that with the change of t_i', 1) agents in the first k layers stays in the original layer, 2) other agents cannot enter the first k layers. So $\mathcal{C}_k$ is independent of t_i'. Furthermore, for each agent in l_k, her valuation of her initial share is the average value of $\mathcal{C}_k$.

Next, we will show that IICM satisfies the incentive compatibility.

Theorem 1. *The Invitation Incentive Cake-cutting Mechanism is incentive compatible.*

Proof. We prove this statement by showing that each agent can maximize her expected utility only when she truthfully reports. At first, we show each agent's utility under the Invitation Incentive Cake-cutting Mechanism.

Given agent i's type t_i, let $(t_i', \mathbf{t}_{-i}')$ be the report profile of all agents. Agent i's final allocation is related to the propagation edges from l_k to l_{k+1}. For each edge in R_k, if it starts from i, agent i get rewards from other agents in the same layer, otherwise part of her initial share will be allocated to others. Then, we consider i's propagation reward shares and the retained parts of her initial share respectively. Let $q = \frac{|\tilde{r}_i((t_i', \mathbf{t}_{-i}'))|}{|R_k((t_i', \mathbf{t}_{-i}'))|}$ represents the propagation contribution of i when report t_i'.

- Propagation reward shares: because of her own propagation, agent i can get $\beta|\tilde{r}_i|$ parts among total $\alpha|R_k|$ parts from each other agent's initial share in l_k. With Lemma 1, i's expected valuation of these shares is
  $$\sum_{j\in l_k\backslash\{i\}} \left(\frac{\beta|\tilde{r}_i|}{\alpha|R_k|} \cdot v_i(X_j)\right) = q\frac{\beta}{\alpha} \cdot v_i\left(\bigcup_{j\in l_k\backslash\{i\}} X_j\right)$$
- Retained initial shares: because of others' propagation, i's initial share will be taken away $(\beta + \gamma)(|R_k| - |\tilde{r}_i|)$ parts among total $\alpha|R_k|$ parts. The retained parts will be finally left to i. The expected valuation of the retained shares is
  $$\frac{\alpha|R_k|-(\beta+\gamma)(|R_k|-|\tilde{r}_i|)}{\alpha|R_k|}v_i(X_i) = \left(1 - \frac{\beta+\gamma}{\alpha} + q\frac{\beta+\gamma}{\alpha}\right)v_i(X_i)$$

With A_i being composed of these two parts and using Lemma 2, agent i's expected utility should be

$$\mathbb{E}(v_i(A_i)) = \left[\frac{1}{|l_k|}\left(1 - \frac{\beta+\gamma}{\alpha}\right) + q(\frac{\beta|l_k|+\gamma}{\alpha|l_k|})\right]v_i(\mathcal{C}_k). \tag{1}$$

Next, we will prove that for each agent i, 1) for all possible reported neighbour set r_i', agent i's dominant strategy is to report her true valuation function, 2) when reporting true valuation function f_i, agent i maximizes her utility if she propagates the information to all her neighbours.

Truthfully Report on Valuation Functions. With regard to the report valuation function f_i', misreporting has no effect on the propagation reward shares, so we only consider the retained shares of i's initial share. The retained shares contain $\left(1 - \frac{\beta+\gamma}{\alpha} + q\frac{\beta+\gamma}{\alpha}\right)$ portion of parts in agent i's initial share. If i reports her true valuation function, then the value of these parts must be $\left(1 - \frac{\beta+\gamma}{\alpha} + q\frac{\beta+\gamma}{\alpha}\right)$. If $f_i' \neq f_i$, this portion may not be achieved. Furthermore, the expected value does not increase along i's misreport, so it is a dominant strategy for i to report her true valuation function f_i.

Truthfully Report on Propagation. In this part, we need to show that each agent can maximize her expected utility when she propagates to all of her neighbours, i.e., reporting $r_i' = r_i$. With all agents' report profile $\mathbf{t}'$, agent i's expected utility is shown in Eq. (1). Inferred from Lemma 2, l_k and $\mathcal{C}_k$ is independent of agent i's report, so there is only one variable q changes with t_i' in Equation (1). Furthermore, q depends only on the propagation graph generated from their diffusion. Therefore, i's expected utility depends only on q, and we prove this statement by proving that each agent can only maximize her expected utility when she truthfully reports her neighbours r_i.

For clear presentation, we use the notations $\tilde{r}_{tru} = \tilde{r}_i(((r_i, f_i), \mathbf{t}_{-i}'))$, $R_{tru} = R_i(((r_i, f_i), \mathbf{t}_{-i}'))$ to represent edge sets when agent i truthfully reports, and $\tilde{r}_{mis} = \tilde{r}_i(((r_i', f_i), \mathbf{t}_{-i}'))$, $R_{mis} = R_i(((r_i', f_i), \mathbf{t}_{-i}'))$ to represent those when agent i misreports the neighbour set. When agent i misreports, some of i's out-edges are removed from the origin propagation network $G((t_i, \mathbf{t}_{-i}'))$. These edges can be divided into the following two types.

1. The end points of the edges are not agents in the next layer. Removing these edges will not change $\tilde{r}_i$ and R_k, thereby it cannot change the allocation to agent i. Hence, we can ignore them.
2. The end points of the edges are agents in the next layer. Removing such an edge will lead to a decrease of 1 in $|\tilde{r}_i|$ and $|R_k|$ respectively. Let δ be the number of such edges when agent i reports r_i'.

Therefore, when agent i misreports, we have $|\tilde{r}_{mis}| = |\tilde{r}_{tru}| - \delta$ and $|R_{mis}| = |R_{tru}| - \delta$ with $\delta \geq 0$. Then there is $\frac{|\tilde{r}_{truth}|}{|R_{truth}|} \geq \frac{|\tilde{r}_{mis}|}{|R_{mis}|}$, which indicates that misreporting neighbours cannot lead to a larger q. Since i's utility monotonically increases with q, i can maximize her utility if she truthfully report.

In the following, we provide an example to illustrate incentive compatibility, as well as to explain how the mechanism parameters affect the allocation results.

Example 1. Suppose the sponsor has 3 neighbours, including agent i, who has $|r_i| = 200$ neighbours in the next layer. With all other agents' report profile $\mathbf{t}_{-i}'$, there are total 50 edges from other agents in l_1 to agents in l_2.

That is, we have $|l_1| = 3$, $|R'_1(((r'_i, f'_i), \mathbf{t}'_{-i}))| = |\tilde{r}'_i(((r'_i, f'_i), \mathbf{t}'_{-i}))| + 50$ and $0 \leq |\tilde{r}'_i(((r'_i, f'_i), \mathbf{t}'_{-i}))| \leq 200$. With Eq. (1), we can find the expected utility of agent i as a function of t'_i, the function curves under different parameters are shown in Fig. 4.

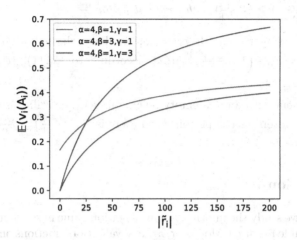

Fig. 4. Relationship between $\mathbb{E}(v_i(A_i))$ and $|\tilde{r}'_i|$ under different parameters when $|l_1| = 3$, $|R'_1| = |\tilde{r}'_i| + 50$ and $0 \leq |\tilde{r}'_i| \leq 200$.

The base case is shown as the curve in Fig. 4 where $\alpha = 4, \beta = 1, \gamma = 1$. The other two curves are the cases where β increases to 3 and γ increases to 3 respectively. The relationship between agent i's valuation of A_i and $|\tilde{r}'_i|$ is shown in Fig. 4. From this figure, we can get the following results.

1. $v_i(A_i)$ monotonically increases with respect to $|\tilde{r}'_i|$, this is consistent with that each agent is incentivized to report all neighbours.
2. When $\beta = 3$, $v_i(A_i)$ increases more dramatically than when $\beta = 1$. The reason is that an agent's propagation reward shares are more relative to her invitation contribution when β increases.
3. When only γ increases, the shares retained from an agent's initial share are less, so the green curve lays below the blue curve. When $|\tilde{r}'_i|/|R'_1| \to 1$, $v_i(A_i)$ is nearly independent of γ, so these two curves are asymptotes.

In addition to incentive compatibility, we also consider fairness among agents. As shown in Proposition 1, we are not able to guarantee proportionality for all agents in an incentive compatible mechanism. However, we consider local fairness within layers instead. This is because our mechanism assigns different priorities to agents in different layers, and the unequal allocation among agents of different priorities should not be considered as unfairness.

Local proportionality requires that every agents in layer l_k receive $1/|l_k|$ part of $\mathcal{C}_k$. We show that our mechanism can approximate such proportionality with a constant ratio. Moreover, the layer proportionality also guarantees agents' incentives to propagate, since each agent's utility is guaranteed no matter whether she propagate the information.

Theorem 2. *In our mechanism, for all $k = 1, 2, \cdots, m$ and all $i \in l_k$, $v_i(A_i)$ satisfies $\mathbb{E}\left(\frac{v_i(A_i)}{v_i(C_k)}\right) \geq \Theta(\frac{1}{|l_k|})$ with constant factor $\frac{\alpha - \beta - \gamma}{\alpha}$.*

Proof. For those agent i with $r_i = \emptyset$, we have that $q = 0$, then there is $\mathbb{E}(v_i(A_i)) = \left[\frac{1}{|l_k|}\left(1 - \frac{\beta + \gamma}{\alpha}\right)\right] v_i(C_k)$ according to Eq. (1). It implies that $\mathbb{E}\left(\frac{v_i(A_i)}{v_i(C_k)}\right) = \frac{\alpha - \beta - \gamma}{\alpha |l_k|}$.

For each agent with an non empty neighbour set, her utility is not less than this because of incentive compatibility, so there is $\mathbb{E}\left(\frac{v_i(A_i)}{v_i(C_k)}\right) \geq \frac{\alpha - \beta - \gamma}{\alpha |l_k|}$ for each agent in l_k.

5   Conclusion

In this paper, we study the information propagation game in social networks with a limited divisible resource. Moreover, agents' valuation functions on the resource are allowed to be different. To maximize information propagation, the sponsor hopes that each agent will propagate the information to all her neighbours. To this end, we propose the Invitation Incentive Cake-cutting Mechanism (IICM) to incentivize agents to invite all their neighbours and report the true valuation to the cake. Furthermore, our mechanism provides a guarantee for minimum share within a propagation layer. Applications of IICM can be developed for social platforms.

Acknowledgement. This work is supported by Science and Technology Commission of Shanghai Municipality (No. 22ZR1442200) and the Shanghai Frontiers Science Center of Human-centered Artificial Intelligence.

References

1. Babaioff, M., Dobzinski, S., Oren, S., Zohar, A.: On bitcoin and red balloons. In: Proceedings of the 13th ACM Conference on Electronic Commerce, EC 2012, pp. 56–73. ACM (2012)
2. Bei, X., Chen, N., Huzhang, G., Tao, B., Wu, J.: Cake cutting: Envy and truth. In: Proceedings of the 26th International Joint Conference on Artificial Intelligence, IJCAI 2017, pp. 3625–3631. ijcai.org (2017)
3. Chen, J., Li, B.: Maximal information propagation via lotteries. In: Feldman, M., Fu, H., Talgam-Cohen, I. (eds.) WINE 2021. LNCS, vol. 13112, pp. 486–503. Springer, Cham (2022). https://doi.org/10.1007/978-3-030-94676-0_27
4. Chen, W., Wang, Y., Yu, D., Zhang, L.: Sybil-proof mechanisms in query incentive networks. In: Proceedings of the 14th ACM Conference on Electronic Commerce, EC 2013, pp. 197–214. ACM (2013)

5. Chen, Y., Lai, J.K., Parkes, D.C., Procaccia, A.D.: Truth, justice, and cake cutting. Games Econ. Behav. **77**(1), 284–297 (2013)
6. Dey, S.: Securing majority-attack in blockchain using machine learning and algorithmic game theory: a proof of work. In: 2018 10th Computer Science and Electronic Engineering Conference, CEEC 2018, pp. 7–10. IEEE (2018)
7. Emek, Y., Karidi, R., Tennenholtz, M.; Zohar, A.: Mechanisms for multi-level marketing. In: Proceedings 12th ACM Conference on Electronic Commerce (EC-2011), pp. 209–218. ACM (2011)
8. Kiang, M.Y., Raghu, T.S., Shang, K.H.: Marketing on the internet - who can benefit from an online marketing approach? Decis. Support Syst. **27**, 383–393 (2000)
9. Kleinberg, J.M., Raghavan, P.: Query incentive networks. In: 46th Annual IEEE Symposium on Foundations of Computer Science (FOCS 2005), 23–25 October 2005, pp. 132–141. IEEE Computer Society (2005)
10. Li, B., Hao, D., Gao, H., Zhao, D.: Diffusion auction design. Artif. Intell. **303**, 103631 (2022)
11. Li, B., Hao, D., Zhao, D., Zhou, T.: Mechanism design in social networks. In: Proceedings of the 31 AAAI Conference on Artificial Intelligence, 4–9 February 2017, San Francisco, California, USA, pp. 586–592. AAAI Press (2017)
12. Liu, Z., et al.: A survey on blockchain: a game theoretical perspective. IEEE Access **7**, 47615–47643 (2019)
13. Mossel, E., Tamuz, O.: Truthful fair division. In: Kontogiannis, S., Koutsoupias, E., Spirakis, P.G. (eds.) SAGT 2010. LNCS, vol. 6386, pp. 288–299. Springer, Heidelberg (2010). https://doi.org/10.1007/978-3-642-16170-4_25
14. Shi, H., Zhang, Y., Si, Z., Wang, L., Zhao, D.: Maximal information propagation with budgets. In: ECAI 2020–24th European Conference on Artificial Intelligence, vol. 325, pp. 211–218. IOS Press (2020)
15. Steihaus, H.: The problem of fair division. Econometrica **16**, 101–104 (1948)
16. Tang, J.C., Cebrián, M., Giacobe, N.A., Kim, H., Kim, T., Wickert, D.B.: Reflecting on the DARPA red balloon challenge. Commun. ACM **54**(4), 78–85 (2011)
17. Woeginger, G.J., Sgall, J.: On the complexity of cake cutting. Discret. Optim. **4**(2), 213–220 (2007)
18. Zhang, Y., Zhang, X., Zhao, D.: Sybil-proof answer querying mechanism. In: Bessiere, C. (ed.) Proceedings of the 29th International Joint Conference on Artificial Intelligence, IJCAI 2020, pp. 422–428. ijcai.org (2020)
19. Zhang, Y., Zhao, D.: Incentives to invite others to form larger coalitions. In: 21st International Conference on Autonomous Agents and Multiagent Systems, AAMAS 2022, pp. 1509–1517 (2022)
20. Zhao, D.: Mechanism design powered by social interactions. In: AAMAS'21: 20th International Conference on Autonomous Agents and Multiagent Systems, pp. 63–67. ACM (2021)
21. Zhao, D.: Mechanism design powered by social interactions: a call to arms. In: Raedt, L.D. (ed.) IJCAI 2022, Vienna, Austria, 23–29 July 2022, pp. 5831–5835. ijcai.org (2022)
22. Zhao, D., Li, B., Xu, J., Hao, D., Jennings, N.R.: Selling multiple items via social networks. In: Proceedings of the 17th International Conference on Autonomous Agents and MultiAgent Systems, AAMAS 2018, pp. 68–76 (2018)

Optimistic Exploration Based on Categorical-DQN for Cooperative Markov Games

Yu Tian[1], Chengwei Zhang[2(✉)] ⓘ, Qing Guo[3], Kangjie Zheng[2], Wanqing Fang[2], Xintian Zhao[2], and Shiqi Zhang[2]

[1] Heilongjiang University of Science and Technology, Harbin, China
[2] Dalian Maritime University, Dalian, China
chenvy@dlmu.edu.cn
[3] Nanyang Technological University, Singapore, Singapore

Abstract. In multiagent reinforcement learning (MARL), independent cooperative learners face numerous challenges when learning the optimal joint policy, such as non-stationarity, stochasticity, and relative over-generalization problems. To achieve multiagent coordination and collaboration, a number of works designed heuristic experience replay mechanisms based on the 'optimistic' principle. However, it is difficult to evaluate the quality of an experience effectively, different treatments of experience may lead to overfitting and be prone to converge to sub-optimal policies. In this paper, we propose a new method named optimistic exploration categorical DQN (OE-CDQN) to apply the 'optimistic' principle to the action exploration process rather than in the network training process, to bias the probability of choosing an action with the frequency of receiving the maximum reward for that action. OE-CDQN is a combination of the 'optimistic' principle and CDQN, using an 'optimistic' re-weight function on the distributional value output of the CDQN network. The effectiveness of OE-CDQN is experimentally demonstrated on two well-designed games, *i.e.*, the CMOTP game and a cooperative version of the boat problem which confronts ILs with all the pathologies mentioned above. Experimental results show that OE-CDQN outperforms state-of-the-art independent cooperative methods in terms of both learned return and algorithm robustness.

Keywords: Cooperative Markov games · Distributional reinforcement learning · Independent learning · Optimistic principle

1 Introduction

Many complex reinforcement learning (RL) tasks, such as multi-robot control [4] and traffic signal control [5,23,24], are often modeled as cooperative multiagent learning problems, where multiple agents work together to learn an optimal joint policy. A natural way to handle coordination tasks is centralized learning of joint actions. Not surprisingly, the majority of relevant works for this problem in recent years are based on centralised learning [8,12,21,22]. However, centralized learning is hard to scale, as the joint action space grows exponentially with the increase of the number

M. Yokoo et al. (Eds.): DAI 2022, LNAI 13824, pp. 60–73, 2023.
https://doi.org/10.1007/978-3-031-25549-6_5

of agents. Besides, in many settings, partial observability and/or communication constraints restrict the learning of agents' policy, which condition only on the local observation of each agent.

To avoid the above two restrictions, works based on independent learning (IL) [17–19] treat other agents as part of the environment and make decisions based on their local observations, actions and rewards only, which are more universally applicable. Different from centralized training methods, cooperative ILs in MARL literature must overcome a rich taxonomy of learning pathologies to converge upon an optimal joint-policy, such as non-stationarity, stochasticity, and relative overgeneralization problems [14,18]. Traditional RL researches on IL MARL literature [11,13,20] mainly based on the **'optimistic'** principle, where agents choose and evaluate an action according to the maximum expected return (MER) or the weighted value of MER and the Expected Return. Such agents optimistically assume that all other agents will act to maximize their rewards. Thus they update the evaluation of action only if (or prefer) the new evaluation is greater than the previous one. While for deep reinforcement learning (DRL) algorithms, it shows natural shortcomings in ILs cooperative problems, such as sample inefficiency, resulting from obsolete experiences being stored inside experience replay memories (ERM) can become inefficient as the policy of other agents change. Thus existing DRL algorithms for ILs all focus on how to identify and discard experience (HDQN [17] and LDQN [19]) or trajectories (IGASIL [9] and NUI-DDQN [18]) based on the 'optimistic' principle, to reduce the probability of miscoordination cause by the learning pathologies mentioned above.

However, the importance of experience (or trajectories) is hard to identify, especially in games with high penalties nearby the optimal joint policy or the optimal policy of the game is much more difficult to be explored than sub-optimal policies. We find that while the above methods deliver promising performances in many tabular Markov games, all of them are prone to suboptimal policies in some complex scenarios, *e.g.*, an environment with continuous state space that simultaneously confronts ILs with some of the mentioned pathologies.

How about applying the 'optimistic' principle to the action execution process rather than in the training process, to bias the probability of choosing the action with the highest return other than the action with the highest expected return? Intuitively, it can reduce the impact of environmental noise occurring in cooperative games with high punishment for an uncoordinated behavior, and meanwhile increase the likelihood of average return being established for policies leading to coordinated outcomes. In addition, it will also increase the quality of experience storied in ERM. The challenge is how to choose an optimistic action from current learned models. For traditional RL methods, such as FMQ and rFMQ [14], they maintain both the ordinary Q-value and the maximum reward value Q_{max}, and then choose actions according to a weighted value combined over the two Q. With the help of latest DRL works, *i.e.*, distributional RL [2,3,6,7,15], we can learn not only the ordinary Q and Q_{max}, but also the whole distribution of the expected return of any state-action pairs. Thus we can design an 'optimistic' action selection strategy based on distributional RL in a higher quality way.

In this work, we propose a new approach called optimistic exploration categorical deep Q Network (OE-CDQN). OE-CDQN lies between the extremes of ordinary Q-value and the maximum reward value Q_{max} like rFMQ, but extends the 'optimistic' principle to MA-DRL based on categorical-DQN (CDQN) [3], which represents the distribution $Z(s, a)$ by the categorical distribution. In detail, OE-CDQN train the expected return distribution $Z(s, a)$ use CDQN, and choose action in a given state according to a refined optimistic average return $Q_\tau(s, a)$, which is averaged by a re-weighted distribution probability $Z_\tau(s, a)$, where τ is a variable that diminishes from 1 to 0 overtime to characterize the degree of optimistic. The 'optimistic' principle is mainly reflected in the design of the re-weight function. Intuitively, the difference between OE-CDQN and other deep RL methods is that OE-CDQN use an 'optimistic' principle in an 'earn good experience' way while others in a 'cut down bad experience' way. We use two specially designed games as the test bed to evaluate OE-CDQN. The first one is CMOTP game, a tabular multiagent cooperative scenario desigend based on the climbing game [14] proposed in papers of LDDQN [19], which contains non-stationarity and stochasticity problems. Based on the climbing game, we design a new environment in a temporally-extended continuous state Markov Game setting that simultaneously confronts ILs with all of the mentioned pathologies, that we call the boat game. Compared with COMTP, the boat game is a better test platform for the reason that: (a) influence degrees of the mentioned pathologies can be controlled arbitrarily by adjusting the river speed and randomness so that we can observe performances of methods in environments with different difficulties; (b) the game has continuous state space, which is more suitable for the test of deep learning based algorithms. Comparing with state-of-the-art works (NUI-DDQN, HDQN, and LDDQN) in the boat game and the COMTP game, OE-CDQN shows better performances in both learning effect and algorithm stability.

2 Definitions and Notations

2.1 Markov Game

We consider the standard multi-agent RL setting, in which the interaction of agents and environment is modeled as a Markov game, where multiple agents make their choices sequentially. Formally, a Markov game is defined by a tuple $G = \langle \mathcal{N}, \mathcal{S}, \mathcal{O}, \mathcal{A}, P, R, \gamma \rangle$. $\mathcal{N}$ is the set of N agents. $\mathcal{S}$ is the set of states and $\mathcal{O} = \langle \mathcal{O}_1, ..., \mathcal{O}_N \rangle$ is the observation set where $\mathcal{O}_i$ is the observation set of agent i. $\mathcal{A} = \langle \mathcal{A}_1, ..., \mathcal{A}_N \rangle$ is the set of the joint actions where $\mathcal{A}_i$ is the set of the action for agent i. $P : S \times \mathcal{A} \times S \rightarrow [0, 1]$ is the transition function returning the probability of transitioning from a state s to s' given a joint action $\langle a_1, ..., a_N \rangle$. $R = \langle r_1, ..., r_N \rangle$ is the reward function where $r_i : S \times \mathcal{A} \rightarrow \mathbb{R}$ specifies the reward for agent i given the state and the joint action. γ is the discount factor. Especially, a Markov game is a team game if every player gets the same reward. Thus, team games are fully cooperative settings, where players have a shared objective.

The policy π_i of agent i represents a mapping from the observation space to a probability distribution over actions: $\pi_i : O_i \rightarrow \Delta(A_i)$, while $\pi = \langle \pi_i, \pi_{-i} \rangle$ refers to a joint policy of all agents and π_{-i} is the joint policies excluding agent i. Given a joint policy π the return (or expected sum of future rewards) for each agent i starting from a state s can be defined by the state-value function (Eq. 1), also known as Q value function, where r_i^t refers to the reward received by agent i at time t:

$$Q_{i,\pi}(s, a) = \mathbb{E}_\pi \left[\sum_{k=0}^{\infty} \gamma^k r_i^{t+k+1} | s_t = s, a_t = a \right] \quad (1)$$

For a Markov game, a joint policy π^* is a Nash equilibrium (NE) if and only if no agent can improve it's gain through unilaterally deviating from π^*. From a group perspective NE are often sub-optimal. In contrast, Pareto-optimality defines a joint policy $\hat{\pi}$ from which no agent can deviate without making at least one other agent worse. A NE joint policy $\hat{\pi}^*$ is Pareto optimal if it is not Pareto-dominated by any other NE. In cooperative multiagent learning literature especially for a team game, convergence to Pareto optimal NE is the most commonly accepted goal to pursue, considering that multiple players cooperate to maximize their goals [14, 19].

2.2 Distributional RL

In distributional RL, the distribution over returns $Z_\pi(s, a)$ is considered instead of the scalar value function $Q_\pi(s, a)$. This change in perspective has yielded new insights into the dynamics of RL [1], and is a useful tool for analysis [10]. Empirically, distributional RL algorithms show improved sample complexity and final performance, as well as increased robustness to hyperparameter variation [2]. An analogous distributional Bellman equation of the form $Z_\pi(s, a) \overset{D}{=} r(s, a) + \gamma Z_\pi(s', a')$ can be derived, where $A \overset{D}{=} B$ denotes that two random variables A and B have equal probability laws, and the random variables s' and a' are distributed according to $P(.|s, a)$ and $\pi(.|s, a)$, respectively. Similar to the scalar setting, a distributional Bellman operator optimality can be defined by

$$\mathcal{T}Z(s, a) : \overset{D}{=} R(s, a) + \gamma Z(s', \arg\max_{a'} \mathbb{E}[Z(s', a')]), \tag{2}$$

There are various choices of the representation of $Z(s, a)$, including the categorical distribution [3], the quantile distribution [7], and the mixture of Gaussian distributions [2]. In this work, we focus on the categorical distribution, *i.e.*, categorical DQN (CDQN) [3], Formally, CDQN models the value distribution using a discrete distribution parametrized by $N \in \mathbb{N}$ and $V_{MIN}, V_{MAX} \in \mathbb{R}$, and whose support is the set of atoms $\{z_i | z_i = V_{MIN} + i\Delta z, 0 \le i < N\}$, $\Delta z := \frac{V_{MAX} - V_{MIN}}{N-1}$. The atom probabilities $p_i(s, a; \theta)$ is given by a neural network θ: $Z_\theta(s, a) = z_i$, w.p., $p_i(s, a; \theta)$.

The value function of (s, a) is defined by $Q(s, a; \theta) = \sum_i z_i p_i(s, a; \theta)$. The network θ is trained by minimizing the loss $l(s, a; \theta)$, which is the cross-entropy term of the KL divergence,

$$l(s, a; \theta) = D_{KL}(\mathcal{T}Z_{\tilde{\theta}}(s, a) || Z_\theta(s, a)), \tag{3}$$

where $\mathcal{T}Z_{\tilde{\theta}}(s, a)$ is the target distribution of (s, a) calculated by Eq. 2 on target network $\tilde{\theta}$.

2.3 Pathologies in Multi-agent ILs

The MA-RL literature provides a rich taxonomy of learning pathologies that cooperative independent learners (ILs) must overcome to converge upon an optimal joint-policy, *e.g.*, relative over-generalisation, non-stationarity, and stochasticity problems [18]. The climbing game (CG) (Fig. 1a) and the partially stochastic climbing game (PSCG) (Fig. 1b) [14] are the most commonly used examples to help explain the pathologies outlined. In CG and PSCG, each agent decides three actions A, B, and C. Both agents receive the same payoff in the matrix corresponding to their joint action. PSCG (Fig. 1b) is an extension of CG, which is the same as CG except that the reward of joint action $\langle B, B \rangle$ is 14 or 0 with equal probability. Both games have two Nash equilibria, *i.e.*, $\langle A, A \rangle$ and $\langle B, B \rangle$. Meanwhile, $\langle A, A \rangle$ is the Pareto-dominate optimal equilibrium.

(a) Climbing Game

		Agent 2		
		A	B	C
	A	11	-30	0
Agent 1	B	-30	7	6
	C	0	0	5

(b) Partially stochastic Climbing Game

		Agent 2		
		A	B	C
	A	11	-30	0
Agent 1	B	-30	14/0	6
	C	0	0	5

Fig. 1. CG and partially stochastic CG

The best strategy of each agent in the CG game depends on the other agent's strategy. Specifically, if initial strategies of an agent is to select an action with equal probability, then action C will be the best choice for the other agent. Further, if an agent chooses C with higher probability, then the other agent's best strategy will be action B, which means that algorithms based on the above assumptions may converge to $\langle B, B \rangle$ with high probability than $\langle A, A \rangle$. **Relative over-generalization** is a type of action shadowing, occurring in games where a sub-optimal NE yields a higher payoff on average when each selected action is paired with an arbitrary action chosen by the other player. Besides, since the other agent is also dynamically adjusting its strategy to optimize its reward, an agent needs to find the optimal strategy in a situation where Markov property is no longer satisfied, which are critical for learning algorithms to guarantee the convergence in single-agent environments. The cooperative problem caused by not meeting Markov properties is named **the non-stationarity problem**. Further, when the reward function is stochastic, the noise in the environment and the behaviors of other agents may both result in the variation of the reward, which makes the source of variation difficult to distinguish, see PSCG game (Fig. 1(b)). The randomness of the environment further increases the difficulty of finding the best strategy for a learning algorithm. This problem is named as **the stochasticity problem**. Algorithms which did not consider these problems may fail to converge to $\langle A, A \rangle$.

3   Optimistic Exploration Categorical Distributional Q Network

In this section, we propose a new method called optimistic exploration categorical distributional Q network (OE-CDQN), an 'optimistic' IL method based on CDQN [3] to addressing the coordination problem in cooperative games. The key idea of OE-CDQN is to optimize the action selection policy when interacting with the environment based on an optimistic exploration strategy, so as to increase the probability that agents can choose the optimal joint actions. Then, CDQN network is trained based on experiences gained from these optimized experiences. Note that the most critical part of the OE-CDQN is the design of the optimistic exploration (OE) strategy, thus we use the simplest distributional RL method CDQN as the distributional representation method of OE-CDQN. The OE strategy can be extended to distributional RL algorithms with more powerful if needed.

3.1 Optimistic Exploration (OE) Strategy

The OE strategy biased the probability of choosing an action of receiving the maximum return for that action based on the 'optimistic' principle, $i.e.$, the assumption that all other agents will act to maximize their reward. Formally, the learning goal of an 'optimistic' agent i in a cooperative task is to find an optimal policy maximizing the optimistically expected gain

$$a_i^* = \max_{a_i} \hat{Q}_i(o_i, a_i) = \max_{a_i} \max_{a_{-i}} Q_i(s, a_i, a_{-i}) \tag{4}$$

where $\hat{Q}_i(o_i, a_i) = \max_{a_{-i}} Q_i(s, a_i, a_{-i})$ is the optimistic estimation of value function of (o, a).

By optimistically assuming that all other agents will act to maximize their reward, all agents choose their optimistic action to interact with the environment. In a deterministic game where the environment has no randomness to state transition and reward function, the 'optimistic' agent fits perfectly and can address the non-stationarity problem. However, this approach leaves agents vulnerable to misleading rewards. To address stochasticity and relative overgeneralization problems in stochastic games, a natural expanding is to learn a goal between the expected gain and optimal gain weighted by the 'degree of optimism'. Thus, action evaluations fluctuate between optimistic and mean evaluations according to the stochasticity of the game. Following this, we design an optimistic exploration strategy based on the distributional represented value function.

Let $\tau \in [0, 1]$ be the variable expressing the 'degree of optimism', where $\tau = 1$ means 'extremely optimistic' in which agent evaluate return of a state-action by the maximum return it obtained. Conversely, agent in $\tau = 0$ evaluates return of a state-action by its expected return, just the same as the original Q. The τ-optimistic Q value function is defined based on the quantile function [16],

$$Q_\tau(o, a) = \frac{1}{1-\tau} \int_{z_\tau}^{+\infty} p(z; o, a) z\, dz, \tag{5}$$

where $p(x; o, a)$ is the probability density function on $\mathbb{R}$ over returns of (o, a) on x. z_τ is the τ quantile of Z, $i.e.$, the value of the inverse function of $F_Z(z; o, a)$ at τ, where $F_Z(z; o, a)$ is the cumulative distribution function of Z, formally,

$$\begin{cases} F_Z(z; o, a) = \mathbb{P}_Z\{x \le z\} = \int_{-\infty}^z p(x; o, a) dx \\ z_\tau = F_Z^{-1}(\tau; o, a) \end{cases} \tag{6}$$

Intuitively, the τ-optimistic Q value $Q_\tau(o, a)$ defined in Eq. 5 is the expectation of the best $1 - \tau$ returns of (o, a). By varying the 'the degree of optimism' variable τ from 1 to 0, we can get return estimations of a given state-action from 'extremely optimistic' to 'extremely non-optimistic', where $Q(o, a)$ is a special case of $Q_\tau(o, a)$ when $\tau = 0$. Then the optimistic exploration strategy of an observation o with degree τ can be defined naturally by selecting the action with the maximal τ-optimistic Q value, *i.e.*,

$$a_\tau = \arg\max_a Q_\tau(o, a). \tag{7}$$

In the learning process of multiagent tasks, agents are exploring at the beginning so that most selected actions are poor choices. Agents with optimistic exploration strategy defined by Eq. 5 choose actions with best returns actively and ignore the bad returns of these actions, to increase the probability of sampling to joint optimal experience. Note that the OE strategy is used in the interaction process, which does not directly affect the training of $Z(o, a)$. As experiences stored in the ERM are treated without distinction, our methods avoid the overestimation problem of actions that appeared in 'optimistic' training methods mentioned above. Once an OE agent has explored, it also needs to choose the average estimation of actions to address the stochasticity problem in stochastic games. So the agents are initially optimistic and the degree of optimism decreases as the time goes on, realized by descrying the optimistic degree τ from 1 to 0.

3.2 Combine OE with Categorical DQN

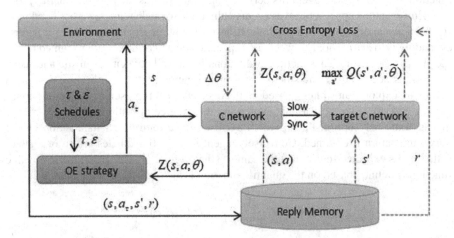

Fig. 2. OECD architecture. We build on the standard Categorical DQN architecture by adding an OE strategy (bottom left). Actions are selected using the ε-greedy exploration method where the greedy part choice actions with maximal $Q_\tau(o, a)$. τ & ε schedules (middle left) decay τ and ε from 1 to 0 gradually.

In this section, we introduce OE-CDQN, an OE strategy algorithm based on CDQN. Here we want to integrate OE strategy into CDQN without changing the original training methods of the algorithm, so that the original performance of CDQN will not be disturbed. State-of-art methods such as HDDQN [17] and LDDQN [19] train the DQN network by modifying the loss and learning rate of different experiences, we will show in experiments that the convergence property of these algorithms is not very good. As shown in Sect. 2.2, CDQN model the value distribution using a discrete distribution on N values $\{z_i\}_{i=1,...,N}$ ($z_i < z_j$ when $i < j$) with probability p_i. Note that the deep network of the CDQN learns the probability distribution $\mathbf{P} = \langle p_1, ..., p_N \rangle$ of each (o, a), we need only to add a mapping function on the output of CDQN network based on Eq. 5 to combine OE strategy with CDQN. Cause $\langle z_i \rangle$ is an ordered monotone increasing sequence, the τ-optimistic Q value function defined in Eq. 5 can be realized as following,

$$Q_\tau(o, a; \theta) = \sum\nolimits_{i=i_\tau}^{N} \hat{p}_i(o, a; \theta) z_i = \frac{\sum_{i=i_\tau}^{N} p_i(o, a; \theta) z_i}{\sum_{i=i_\tau}^{N} p_i(o, a; \theta)} \qquad (8)$$

where i_τ is the index of the τ quantile, formally,

$$i_\tau = \min i, \quad \text{s.t.} \quad \sum\nolimits_{k=1}^{i} p_k(o, a; \theta) > \tau \qquad (9)$$

The optimization problem in Eq. 9 needs no more than one traverse process, *i.e.*, i_τ can be found by adding p_k from 1 to N until the sum is greater than τ, so the application of OE strategy to CDQN will not significantly increase the computational complexity.

Figure 2 shows the learning framework of OE-CDQN. The right half of the framework is the training part of OE-CDQN, which remains the same as CDQN. In the training process, agents sample a batch of experiences from ERM randomly and update the CDQN network θ by minimizing the cross-entropy loss (Eq. 3). The target CDQN network $\tilde{\theta}$ is synchronized with the CDQN network θ at a slightly slower speed. The main difference between OE-CDQN and CDQN is the interaction process between agent and environment as shown in the left side of the framework. CDQN chooses action to interact according to the ϵ-greedy exploration method, *i.e.*, with probability ϵ select a random action, otherwise select the action with the maximal Q value, while OE-CDQN selects the action using the τ-optimistic Q, *i.e.*, the greedy part choice actions with maximal $Q_\tau(o, a)$. τ & ϵ schedules (middle left) update τ and ϵ from 1 to 0 gradually as the number of interactions increases.

4 Experimental Results

4.1 Game Description

We use two specially designed games as the test bed to evaluate OE-CDQN, *i.e.*, the CMOTP game (Fig. 3(a)) and the boat game (Fig. 3(b)). Specifically, CMOTP is a Markov game extension of the Climbing game [14] proposed in LDDQN [19], in which two agents are tasked with delivering one item of goods to drop zones within a grid-world cooperatively. Multiple target zone and stochastic rewards make CMOTP

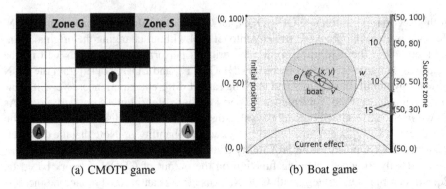

(a) CMOTP game   (b) Boat game

Fig. 3. (a) CMOTP game, Zone G yields a reward of 8, whereas drop Zone S returns a reward of 12 on 50% of occasions and only 0 on the other 50%. (b) Boat game, the three quay in the right side of the river divides the success zone into three parts: '−' optimal zone; ' ' sub-optimal zone; and '−' bad zone.

suffer from non-stationarity and stochasticity pathologies. Besides CMOTP, we further design a temporally-extended continuous state Markov game denoted as the boat game. Compared with CMOTP, this new game is more challenging since it confronts ILs with the three pathologies introduced in Sect. 2.3, *i.e.*, *relative over-generalization, the non-stationarity problem*, and *stochasticity problem*.

Settings of the Boat Game. The goal of our boat game is to control the forward acceleration and the angular acceleration of the boat by two independent agents to optimize expected return. State space of the game is defined by the state of the boat and the speed of water; Observation of the two agents is defined the same as state except for the water speed e. The speed of water is defined by $e = f_t[x/l - (x/l)^2]$, where l is the width of the river, x is the horizontal coordinate of the boat, f_t is the current force subject to a normal distribution $N(\mu, \sigma^2)$ at time t. *The random current force makes the game suffers from* ***the stochasticity problem*** *where we would receive random rewards.* We can control the difficulty degree of this pathology by setting different noise, *i.e.*, the std σ. For the reward function R, we define three target quays (a sub-optimal zones, an optimal zone, and a bad zone) at the right side of the river, where different reward will be given when the boat reaches different zone. Ranges of the two sub-optimal zone are large and easy to explore, while the optimal range is small and easy to be affected by environmental randomness and actions of other agents. *The dynamic optimal range makes the game suffer from* ***the relative overgeneralization problem***. Besides, rewards are set to −0.1 before the boat reaches the destination (the right side of the river) at each time of an episode, to guide the boat through the river quickly. Influenced by the water flow, it takes more steps for reaching the optimal zone (about 35 steps) than the sub-optimal zone (about 25 steps), resulting in best returns of the two zones equal to 11.5 and 7.5, respectively. This setting further increases the difficulty of the pathology. *Since each agent knows only parts of state information, and do not communicate with each other, the game also suffers from the* ***the non-stationarity problem***.

4.2   Experiment Settings

Baselines. We compare OE-CDQN with the state-of-the-art independent MARL methods, *i.e.*, HDQN [17], LDDQN [19], and NUI-DDQN [18]. Note that, LDDQN calculates leniency of each state action pair according to their frequency, resulting a poor performance in the game with continuous state. Thus, we modify LDDQN by redesigning its leniency strategy and denote this new version as LDDQN-M. Specifically, we make following modifications for LDDQN: 1) we set the leniency of the LDDQN linearly decrease over time, which is similar to ϵ-greedy decay strategy. 2) we add weights to the experience stored in ERM of LDDQN according to its storage time to reduce the importance of earlier experiences. Unless stated otherwise, all networks use the same architecture and hyper-parameters. The parameters of all algorithm are detailed in the supplementary material.

Experimental Setups. We conduct three experiments on the boat game: (a) *Learning efficiency and robustness comparison*. All compared methods are evaluated under different noise intensity (*i.e.*, noise $\sigma = 0$, 0.5 and 1.0); (b) *Hyper-parameter analysis*. We conduct comprehensive experiments according to different hyper-parameter settings and detect the valid range of these parameters (*i.e.*, optimistic decay rate δ_τ and categorical number N); (c) *Ablation study*. We conduct a comparison between our OE-CDQN and CDQN to validate the necessity and effectiveness of the OE strategy. (d) Validation on CMOTP game. We further conduct the learning efficiency comparisons to show that OE-CDQN can run well on the existing test bed.

4.3   Results in CMOTP Game

We compared OE-CDQN with LDDQN, HDQN and NUI-DDQN in CMOTP game. We refined the return of each methods by minus the length of the episode times 0.001 as the measure of learning results,*i.e.*, how fast the two agents learn the joint optimal policy. From Fig. 4, we can conclude that our algorithm also learns the optimal strategy stably in the CMOTP game. Methods based on optimistic training, *i.e.*, LDDQN, HDQN and NUI-DDQN, are very prone to overestimation, and we didn't find an effective set of parameters to achieve a desired results for those methods. All of those method mis-cooperated after a period of training.

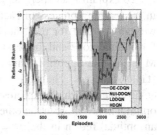

Fig. 4. Learning performances in CMOTP game.

4.4   Results and Analysis in the Boat Game

Learning Efficiency and Robustness. We compare OE-CDQN with methods mentioned above in the boat game with noise $\sigma = 0$, 0.5, and 1 respectively. As depicted in Fig. 5, only OE-CDQN and LDDQN-M learned a relatively stable strategy, with OE-CDQN reached the global optimal return (nearby 11) and LDDQN-M reached the suboptimal (nearby 7). Besides, it can be seen from the three pictures that the environmental

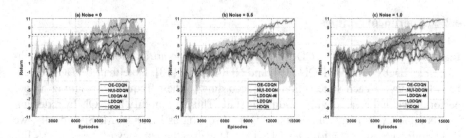

Fig. 5. Learning performance of OE-CDQN compared with HDQN, LDQN, LDQN-M and NUI-DQN under different noise intensity: (a) noise $\sigma = 0$, (b) $\sigma = 0.5$ and (c) $\sigma = 1.0$

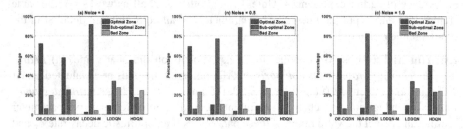

Fig. 6. Statistics results of OE-CDQN compared with HDQN, LDQN, LDQN-M and NUI-DQN under different noise intensity: (a) noise $\sigma = 0$, (b) $\sigma = 0.5$ and (c) $\sigma = 1.0$

noise ranges from 0 to 1 has little effect on OE-CQDN, which means the robustness of the algorithm to the environmental noise is acceptable.

To explain the experiment results, we calculated the final location of each episode in the learning process of each algorithm, shown in Fig. 6. From the statistical results, we can see that for each noise settings, OE-CDQN gets the most exploration times for optimal zone and the least exploration times for sub-optimal zone, while LDDQN-M is just the opposite. This intuitively explains why these two algorithms can learn stable strategies. The statistical results of NUI-DDQN vary greatly in different noise environments, which explores the optimal zone most in noise $\sigma = 0$ and explores sub-optimal zones most in the other two environments. The main reason why it didn't learn an optimal return in the $\sigma = 0$ is that the explorations number of sub-optimal zone of NUI-DDQN is relatively high, resulting in the relative overgeneralization problem which interfere with the estimation of the algorithm. For LDDQN and HDDQN, the optimistic training of experience makes them overestimate some bad strategies, thus lots of trail of them reach the sub-optimal zone.

Adaptation of Parameters and Ablation Study. We discuss the application scope of the two unique parameters in OE-CDQN, $i.e.$, the optimistic decay rate d_τ, and the atom number of CDQN network N_a. Figure 7(a) and (b) show the learning performance of OE-CDQN with four different optimistic decay rates (0.65, 0.70, 0.75 and 0.80 times of ε decay rate d) and atom numbers of CDQN network (40, 45, 51 and 60) in the boat game with noise $\sigma = 1$ respectively.

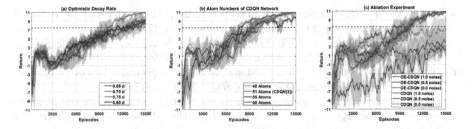

Fig. 7. Learning performance of OE-CDQN with different parameter setting (a & b); Ablation study (c)

It can be seen that all tests have learned more than 9 returns. Note that the highest return of the sub-optimal zone is about 7.5, which means all the tests have learned the optimal zone. We can obtain that OE-CDQN is fairly stable when the change is moderate. We also compared OE-CDQN with CDQN to show the necessity of the proposed OE strategy. Figure 7(c) shows the learning performance of OE-CDQN and CDQN in the boat game with all noise settings. We can see that CDQN can't solve this problem alone.

5 Conclusion

In this work, we have proposed the optimistic exploration based categorical DQN (OE-CDQN) by introducing the 'optimistic' exploration strategy into categorical distributional reinforcement learning (RL). Our method realized 'optimistic' principle based action exploration, which can address the coordination problem in cooperative games effectively, e.g., the pathologies of independent learning cooperative Markov games. We validate and compare our method and three state-of-art methods on a public available CMOTP game and a more challenging and self-designed boat game. The experimental results showed that our method outperforms state-of-the-art methods in both learned return and algorithm robustness.

Acknowledgment. The work is supported by the National Natural Science Foundation of China (Grant Nos.:61906027).

References

1. Azar, M.G., Munos, R., Kappen, B.: On the sample complexity of reinforcement learning with a generative model. In: Proceedings of the 29th International Conference on Machine Learning, ICML 2012 (2012)
2. Barth-Maron, G., et al.: Distributed distributional deterministic policy gradients. In: 6th International Conference on Learning Representations, ICLR 2018 (2018)
3. Bellemare, M.G., Dabney, W., Munos, R.: A distributional perspective on reinforcement learning. In: Proceedings of the 34th International Conference on Machine Learning, ICML 2017 (2017)

4. Cao, Y., Yu, W., Ren, W., Chen, G.: An overview of recent progress in the study of distributed multi-agent coordination. IEEE Trans. Industr. Inform. **9**(1), 427–438 (2013)
5. Chu, T., Wang, J., Codecà, L., Li, Z.: Multi-agent deep reinforcement learning for large-scale traffic signal control. IEEE Trans. Intell. Transp. Syst. **21**, 1086–1095 (2019)
6. Dabney, W., Ostrovski, G., Silver, D., Munos, R.: Implicit quantile networks for distributional reinforcement learning. arXiv preprint arXiv:1806.06923 (2018)
7. Dabney, W., Rowland, M., Bellemare, M.G., Munos, R.: Distributional reinforcement learning with quantile regression. In: Proceedings of the Thirty-Second AAAI Conference on Artificial Intelligence, AAAI 2018) (2018)
8. Foerster, J.N., Farquhar, G., Afouras, T., Nardelli, N., Whiteson, S.: Counterfactual multi-agent policy gradients. In: Thirty-Second AAAI Conference on Artificial Intelligence (2018)
9. Hao, X., Wang, W., Hao, J., Yang, Y.: Independent generative adversarial self-imitation learning in cooperative multiagent systems. In: Proceedings of the 18th International Conference on Autonomous Agents and MultiAgent Systems, AAMAS 2019 (2019)
10. Lattimore, T., Hutter, M.: PAC bounds for discounted MDPs. In: Bshouty, N.H., Stoltz, G., Vayatis, N., Zeugmann, T. (eds.) ALT 2012. LNCS (LNAI), vol. 7568, pp. 320–334. Springer, Heidelberg (2012). https://doi.org/10.1007/978-3-642-34106-9_26
11. Lauer, M., Riedmiller, M.A.: An algorithm for distributed reinforcement learning in cooperative multi-agent systems. In: Proceedings of the 17h International Conference on Machine Learning, pp. 535–542 (2000)
12. Lowe, R., WU, Y., Tamar, A., Harb, J., Pieter Abbeel, O., Mordatch, I.: Multi-agent actor-critic for mixed cooperative-competitive environments. In: Advances in Neural Information Processing Systems, vol. 30, pp. 6379–6390. Curran Associates, Inc. (2017)
13. Matignon, L., Laurent, G.J., Fort-Piat, N.L.: Hysteretic Q-learning: an algorithm for decentralized reinforcement learning in cooperative multi-agent teams. In: IEEE/RSJ International Conference on Intelligent Robots and Systems IROS, pp. 64–69. IEEE (2007)
14. Matignon, L., Laurent, G.J., Le Fort Piat, N.: Review: independent reinforcement learners in cooperative Markov games: a survey regarding coordination problems. Knowl. Eng. Rev. **27**(1), 1–31 (2012)
15. Mavrin, B., Yao, H., Kong, L., Wu, K., Yu, Y.: Distributional reinforcement learning for efficient exploration. In: International Conference on Machine Learning, pp. 4424–4434 (2019)
16. Müller, A.: Integral probability metrics and their generating classes of functions. Adv. Appl. Probab. **29**(2), 429–443 (1997)
17. Omidshafiei, S., Pazis, J., Amato, C., How, J.P., Vian, J.: Deep decentralized multi-task multi-agent reinforcement learning under partial observability. In: Proceedings of the 34th International Conference on Machine Learning, vol. 70, pp. 2681–2690. JMLR. org (2017)
18. Palmer, G., Savani, R., Tuyls, K.: Negative update intervals in deep multi-agent reinforcement learning. In: Proceedings of the 18th International Conference on Autonomous Agents and MultiAgent Systems (2019)
19. Palmer, G., Tuyls, K., Bloembergen, D., Savani, R.: Lenient multi-agent deep reinforcement learning. In: Proceedings of the 17th International Conference on Autonomous Agents and MultiAgent Systems (2018)
20. Panait, L., Sullivan, K., Luke, S.: Lenient learners in cooperative multiagent systems. In: Proceedings of the 5th International Joint conference on Autonomous Agents and Multiagent Systems (2006)
21. Rashid, T., Samvelyan, M., Witt, C.S., Farquhar, G., Foerster, J., Whiteson, S.: QMIX: monotonic value function factorisation for deep multi-agent reinforcement learning. In: International Conference on Machine Learning, pp. 4292–4301 (2018)

22. Son, K., Kim, D., Kang, W.J., Hostallero, D., Yi, Y.: QTRAN: learning to factorize with transformation for cooperative multi-agent reinforcement learning. In: Proceedings of the 36th International Conference on Machine Learning, ICML 2019 (2019)
23. Zhang, C., Jin, S., Xue, W., Xie, X., Chen, S., Chen, R.: Independent reinforcement learning for weakly cooperative multiagent traffic control problem. IEEE Trans. Veh. Technol. **70**(8), 7426–7436 (2021)
24. Zhang, C., et al.: Neighborhood cooperative multiagent reinforcement learning for adaptive traffic signal control in epidemic regions. IEEE Trans. Intell. Transp. Syst. 1–12 (2022)

A Game-Theoretic Approach
to Multi-agent Trust Region Optimization

Ying Wen[1]([✉]), Hui Chen[2], Yaodong Yang[3], Minne Li[2], Zheng Tian[4],
Xu Chen[5], and Jun Wang[2]

[1] Shanghai Jiao Tong University, Shanghai, China
ying.wen@sjtu.edu.cn
[2] University College London, London, UK
[3] Peking University, Beijing, China
[4] ShangahiTech University, Shanghai, China
[5] Renmin University, Beijing, China

Abstract. Trust region methods are widely applied in single-agent reinforcement learning problems due to their monotonic performance-improvement guarantee at every iteration. Nonetheless, when applied in multi-agent settings, the guarantee of trust region methods no longer holds because an agent's payoff is also affected by other agents' adaptive behaviors. To tackle this problem, we conduct a game-theoretical analysis in the policy space, and propose a multi-agent trust region learning method (MATRL), which enables trust region optimization for multi-agent learning. Specifically, MATRL finds a stable improvement direction that is guided by the solution concept of Nash equilibrium at the meta-game level. We derive the monotonic improvement guarantee in multi-agent settings and show the local convergence of MATRL to stable fixed points in differential games. To test our method, we evaluate MATRL in both discrete and continuous multiplayer general-sum games including checker and switch grid worlds, multi-agent MuJoCo, and Atari games. Results suggest that MATRL significantly outperforms strong multi-agent reinforcement learning baselines.

Keywords: Multi-agent Reinforcement Learning · Game Theory · Trust Region Optimization

1 Introduction

Multi-agent systems (MASs) [29] have received much attention from the reinforcement learning community [40]. In the real world, automated driving [43], StarCraft II [25,37] and Dota 2 [3] are a few examples of the myriad of applications that can be modeled by MASs. Due to the complexity of multi-agent problems, an investigation into whether agents can learn to behave effectively during interactions with environments and other agents is essential [10]. This investigation can be conducted naively through an *independent learner* (IL) [32], which

M. Yokoo et al. (Eds.): DAI 2022, LNAI 13824, pp. 74–87, 2023.
https://doi.org/10.1007/978-3-031-25549-6_6

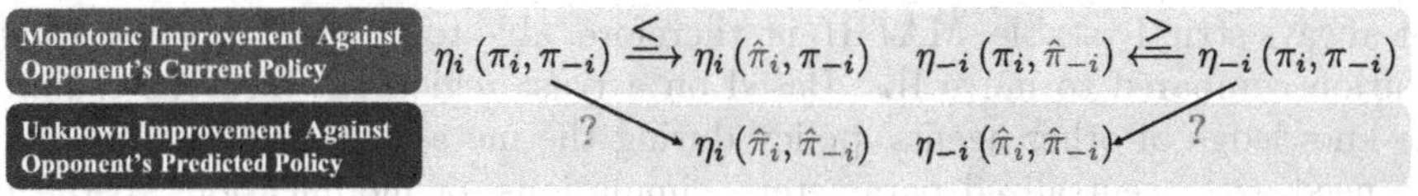

Fig. 1. Discounted returns η_i for an agent i given different joint policy pairs, where π_i is the current policy, and π'_i is the simultaneously predicted policy. Given π_i, the monotonic improvements of a fixed opponent can be easily measured: $\eta_i(\pi'_i, \pi_{-i}) \geq \eta_i(\pi_i, \pi_{-i})$. However, due to simultaneous learning, the improvement of $\eta_i(\pi'_i, \pi'_{-i})$ is unknown compared to $\eta_i(\pi_i, \pi_{-i})$.

ignores the other agents and optimizes the policy assuming a stable environment [5]; and *trust region* method (e.g., proximal policy optimization (PPO) [28]) based ILs are popular [37] due to their theoretical guarantee for single-agent learning and good empirical performance in real-world applications.

In multi-agent scenarios, however, an agent's improvement is affected by other agents' adaptive behaviors (i.e., the multi-agent environment is *nonstationary* [11]). As a result, trust region learners can measure the policy improvements of agents' predicted policies compared to the current policies, but the improvements compared to the other agents' predicted policies are still unknown (shown in Fig. 1). Therefore, trust-region-based ILs perform worse in MASs than in single-agent tasks. Moreover, the convergence to a *fixed point*, such as a *Nash equilibrium* [4], is a common and widely accepted solution concept for multi-agent learning. Thus, although ILs can best respond to other agents' current policies, they lose their convergence guarantee.

One solution for addressing the convergence problem for ILs is empirical game-theoretic analysis (EGTA) [38], which approximates the best response to the policies generated by ILs [23]. Although EGTA-based methods [1,14,24] establish convergence guarantees in several game classes, their computational cost is also large when empirically approximating and solving the meta-game. Other multi-agent learning approaches collect or approximate additional information such as communication [25] and centralized joint critics [19]. Nevertheless, these methods usually require centralized critics or centralized communication assumptions, which require extra training efforts. Thus, there is considerable interest in the use of multi-agent learning to find an algorithm that while having minimal requirements and computational cost as ILs, also simultaneously improves convergence performance.

This paper presents a *multi-agent trust region learning* (MATRL) algorithm that augments the trust region ILs with a meta-game analysis to improve learning stability and efficiency. In MATRL, a trust region trial step for an agent's payoff improvement is implemented by ILs, which provide a predicted policy based on the current policy. Then, an empirical policy-space meta-game is constructed to compare the expected advantages of the predicted policies with those of the current policies. By solving the meta-game, MATRL finds a restricted step by aggregating the current and predicted policies using the meta-game Nash equilibrium. Finally, MATRL takes the best responses based on the aggregated policies from the last step for each agent to explore because the identified stable trust region is not always strictly stable. MATRL is, therefore, able to provide a weakly stable solution compared to naive ILs. Based on a trust region IL, MATRL requires the knowledge of other agents' policy during the meta-game analysis but does not need extra centralized parameters, simulations, or modifications to the IL itself. We provide insights into the empirical meta-game in Sect. 2.2, showing that the approximated Nash equilibrium of the meta-game is a weak stable fixed point of the underlying game. Our experiments demonstrate that MATRL significantly outperforms deep ILs [28] with the same hyperparameters, VDN [31], QMIX [26] and QDPP [41] methods in discrete action grid worlds, decentralized PPO ILs, centralized MADDPG [19] and independent DDPG and COMIX [39] in a continuous action multi-agent MuJoCo task [39] and zero-sum multi-agent Atari [34].

2 Multi-agent Trust Region Learning

Notations and Preliminaries. A stochastic game [18] can be defined as follows: $\mathcal{G} = \langle \mathcal{N}, \mathcal{S}, \{\mathcal{A}_i\}, \{\mathcal{R}_i\}, \mathcal{P}, p_0, \gamma \rangle$, where $\mathcal{N}$ is a set of agents, $n = |\mathcal{N}|$ is the number of agents, and $\mathcal{S}$ denotes the state space. $\mathcal{A}_i$ is the action space for agent i. $\mathcal{A} = \mathcal{A}_1 \times \cdots \times \mathcal{A}_n = \mathcal{A}_i \times \mathcal{A}_{-i}$ is the joint action space, and for simplicity, we use $-i$ to denote agents other than agent i. $\mathcal{R}_i = R_i(s, a_i, a_{-i})$ is the reward function for agent $i \in \mathcal{N}$. $\mathcal{P} : \mathcal{S} \times \mathcal{A} \times \mathcal{S} \rightarrow [0,1]$ is the transition function. p_0 is the initial state distribution, and $\gamma \in [0,1)$ is a discount factor. Each agent $i \in \mathcal{N}$ has a stochastic policy $\pi_i(a_i|s) : \mathcal{S} \times \mathcal{A}_i \rightarrow [0,1]$ and aims to maximize its long-term discounted return:

$$\eta_i(\pi_i, \pi_{-i}) = \mathbb{E}_{s^0, a_i^0, a_{-i}^0 \cdots} \left[\sum_{t=0}^{\infty} \gamma^t R_i(s^t, a_i^t, a_{-i}^t) \right], \tag{1}$$

where $s^0 \sim p_0$, $s^{t+1} \sim \mathcal{P}(s^{t+1}|s^t, a_i^t, a_{-i}^t)$, and $a_i^t \sim \pi_i(a_i^t|\tau_i^t)$.

Then, we have the standard definitions of the state-action value and state value functions: $Q_i^{\pi_i, \pi_{-i}}(s^t, a_i^t, a_{-i}^t) = \mathbb{E}_{s^{t+1}, a_i^{t+1}, a_{-i}^{t+1} \cdots}[\sum_{l=0}^{\infty} \gamma^l R_i(s^{t+l}, a_i^{t+l}, a_{-i}^{t+l})]$ and $V_i^{\pi_i, \pi_{-i}}(s^t) = \mathbb{E}_{a_i^t, a_{-i}^t, s^{t+1} \cdots}[\sum_{l=0}^{\infty} \gamma^l R_i(s^{t+l}, a_i^{t+l}, a_{-i}^{t+l})]$; also the advantage function $A_i^{\pi_i, \pi_{-i}}(s^t, a_i^t, a_{-i}^t) = Q_i^{\pi_i, \pi_{-i}}(s^t, a_i^t, a_{-i}^t) - V_i^{\pi_i, \pi_{-i}}(s^t)$, given the state and joint action.

Motivations. A trust region algorithm aims to answer two questions: how to compute a trial step and whether the trial step should be accepted. In multi-agent learning, a trial step toward agents payoff improvement can be easily implemented with ILs, denoted as *independent improvement direction (IID)*. The remaining issue is resolved by finding a restricted step leading to a stable improvement direction, which is not in the single agent's policy space but in the joint policy space. In other words, MATRL decomposes trust region learning into two parts: first, an IID between *current policy* π_i and *predicted policy* $\hat{\pi}_i$ should be identified; then, with the help of the predicted policy, a more refined method, to some extent, can approximate a stable trial step. Instead of line searching in

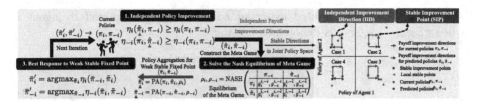

Fig. 2. (Left): Overview of the MATRL phases. The pale red area indicates independent payoff improvement directions; the pale blue area shows stable improvement directions in joint policy space and π: current policy, $\hat{\pi}$: predicted policy in IID step, $\bar{\pi}$: aggregated policy in SIP step; π', next policy. **(Right):** the gray area illustrates IID and SIP with a two-agent game, in which the arrows indicate the payoff improvement directions for agents. The IID guarantees the partially monotone game in red arrows; then, the SIPs are determined by improvement directions (include four cases) of predicted policies in blue arrows.

a single-agent payoff improvement direction, MATRL searches for a joint policy space to achieve a conservative and stable improvement. Essentially, MATRL is an extension of the single-agent TRPO to a MAS, which learns to find a stable point between the current policy and the predicted policy. To find the stable improvement directions, we assume knowledge about other agents' policies during training to avoid unstable improvement via empirical meta-game analysis, while the execution can still be fully decentralized. We explain every step of MATRL in detail in the following sections (also in Fig. 2).

2.1 Independent Trust Payoff Improvement

Single-agent reinforcement learning algorithms can be straightforwardly applied to multi-agent learning, where we assume that all agents behave independently. In this section, we have chosen the policy-based reinforcement learning method— ILs. In multi-agent games, the environment becomes a Markov decision process for agent i when each of the other agents plays according to a fixed policy. We set agent i to make a monotonic improvement against its opponents' fixed policies. Thus, at each iteration, the policy is updated by maximizing the utility function η_i over a local neighborhood of the current joint policy π_i, π_{-i}. We can adopt TRPO (or, PPO [28]), which constrains the step size in the policy update:

$$\hat{\pi}_i = \arg\max_{\pi \in \Pi_{\theta_i}} \eta_i(\pi, \pi_{-i}) \quad \text{s.t. } D\left(\pi_i, \hat{\pi}_i\right) \leq \delta_i, \tag{2}$$

where D is a distance measurement, and δ_i is a constant. Independent trust region learners produce the monotonically improved policy $\hat{\pi}_i$, which guarantees $\eta_i\left(\hat{\pi}_i, \pi_{-i}\right) \geq \eta_i\left(\pi_i, \pi_{-i}\right)$ and provides a trust payoff bound by $\hat{\pi}_i$. Due to simultaneous policy improvement without awareness of other agents , however, the lower bound of payoff improvement from single-agent [27] no longer holds for multi-agent payoff improvement. By following a similar logic in proof, we can obtain a precise lower bound for a simultaneous-move multi-agent payoff improvement.

Remark 1. The approximated expected advantage $g_i^{\pi_i,\pi_{-i}}$ gained by agent i when $\pi_i, \pi_{-i} \rightarrow \hat{\pi}_i, \hat{\pi}_{-i}$ is denoted as follows:

$$g_i^{\pi_i,\pi_{-i}}(\hat{\pi}_i, \hat{\pi}_{-i}) := \sum_s p^{\pi_i,\pi_{-i}}(s) \sum_{a_i,a_{-i}} \hat{\pi}_i(a_i|s)\hat{\pi}_{-i}(a_{-i}|s)A_i^{\pi_i,\pi_{-i}}(s,a_i,a_{-i}),$$

$$(3)$$

where $p^{\pi_i,\pi_{-i}}(s)$ discounted state visitation frequencies induced by π_i, π_{-i}. Then, the following lower bound can be derived for multi-agent independent trust region optimization:

$$\eta_i(\hat{\pi}_i, \hat{\pi}_{-i}) - \eta_i(\pi_i, \pi_{-i}) \geq g_i^{\pi_i,\pi_{-i}}(\hat{\pi}_i, \hat{\pi}_{-i}) - \frac{4\gamma\epsilon_i}{(1-\gamma)^2}(\alpha_i + \alpha_{-i} - \alpha_i\alpha_{-i})^2, \quad (4)$$

where $\epsilon_i = \max_{s,a_{-i},a_{-i}} |A_i^{\pi_i,\pi_{-i}}(s,a_i,a_{-i})|$, $\alpha_i = \max_s D_{\mathrm{TV}}(\pi_i(\cdot|s)\|\hat{\pi}_i(\cdot|s))$ for agent i, and D_{TV} is the total variation divergence.

Based on the independent trust payoff improvement, although the predicted policy $\hat{\pi}_i$ will guide us in determining the step size of the IID, the stability of $(\hat{\pi}_i, \hat{\pi}_{-i})$ is still unknown. As shown in Remark 1, an agent's lower bound is approximately $O(4\alpha^2)$, which is four times larger than the single-agent lower bound trust region of $O(\alpha^2)$. Furthermore, $\epsilon_i = \max_{s,a_{-i},a_{-i}} |A_i^{\pi_i,\pi_{-i}}(s,a_i,a_{-i})|$ depends on other agents' action a_{-i}, which will be very large when agents have conflicting interests. Therefore, the most critical issue underlying MATRL is finding a Stable Improvement Point (SIP) after the IID. In the next section, we illustrate how to search for a weak stable fixed point within the IID based on the meta-game analysis.

2.2   Approximating the Weak Stable Fixed Point

Stabilizing the independent trust payoff improvements is one of the essential components of MATRL. Since each iteration of MATRL requires the solving of additional stable improvement subproblem, finding an efficient solver for this subproblem is very important. Instead of using the stable fixed points [2] as the stable improvement target, we choose the *weak stable fixed point* in Definition 2, which is easier to find. To maximize the objective defined in Eq. (1), we can ask that *reasonable* algorithms avoid all strict minimums, which imposes only that agents are well-behaved regarding strict minima, even if their individual behaviors are not self-interested. Before providing the clear definitions for these points, we first define a differentiable game restricted by the IID:

Definition 1 (Differentiable Restricted Game (DRG)). *If the policy space for each agent i in a game is restricted to open sets $\bar{\Pi}_i = [\pi_i, \hat{\pi}_i] \subseteq \Pi_i$, where $\bar{\Pi}_i \subseteq \Pi_i$, and the expected advantage g_i is twice continuously differentiable in this range, then we call it a differentiable restricted game.*

Denote the simultaneous gradient of the DRG as $\boldsymbol{\xi}(\pi_i, \pi_{-i}) = (\nabla_{\pi_i} g_i, \nabla_{\pi_{-i}} g_{-i})$. We introduce the Hessian of DRG as the block matrix $H = \nabla_{\pi_i,\pi_{-i}}\boldsymbol{\xi}(\pi_i, \pi_{-i})$ to define the types of fixed points:

Definition 2 (Weak Stable Fixed Point). *A point $(\bar{\pi}_i, \bar{\pi}_{-i})$ is a fixed point if $\boldsymbol{\xi}(\bar{\pi}_i, \bar{\pi}_{-i}) = \mathbf{0}$. We then say that $(\bar{\pi}_i, \bar{\pi}_{-i})$ is stable if $H(\bar{\pi}_i, \bar{\pi}_{-i}) \preceq 0$, is unstable if $H(\bar{\pi}_i, \bar{\pi}_{-i}) \succ 0$ and is a* **weak stable fixed point** *if $H(\bar{\pi}_i, \bar{\pi}_{-i}) \not\succ 0^1$.*

We denote the weak stable fixed points in the DRG as the **stable improvement point (SIP)**, it is reasonable if it converges only to fixed points and avoids unstable fixed points (strict minimum) almost completely. Given that we already have the IID, which produces a predicted policy, with the knowledge about all agents policies, it is natural to conduct an EGTA [36] to search for a SIP in the area bounded by the current and predicted policy pair. We then define a meta-game in which each agent i has only two strategies $\pi_i, \hat{\pi}_i$:

$$
\mathcal{M}(\pi_i, \hat{\pi}_i, \pi_{-i}, \hat{\pi}_{-i}) = \begin{pmatrix} g_i^{i,-i}, g_{-i}^{i,-i} & g_i^{i,-\hat{i}}, g_{-i}^{i,-\hat{i}} \\ g_i^{\hat{i},-i}, g_{-i}^{\hat{i},-i} & g_i^{\hat{i},-\hat{i}}, g_{-i}^{\hat{i},-\hat{i}} \end{pmatrix}, \tag{5}
$$

where $g_i^{\hat{i},-\hat{i}} = g_i^{\pi_i, \pi_{-i}}(\hat{\pi}_i, \hat{\pi}_{-i})$ (as defined in Eq. (3)) is an empirical payoff entry of the meta-game, and note that $g_i^{i,-i} = 0$, as it has an expected advantage over itself. Compared with using $\eta_i(\hat{\pi}_i, \hat{\pi}_{-i}) = \eta_i(\pi_i, \pi_{-i}) + g_i^{\hat{i},-\hat{i}}$ as the meta-game payoff, $g_i^{\hat{i},-\hat{i}}$ has lower variance and is easier to approximate because $\eta_i(\pi_i, \pi_{-i})$ is a constant baseline. However, most entries in $\mathcal{M}$ are unknown, and many extra simulations are required to estimate the payoff entries (e.g., $g_i^{\hat{i},-\hat{i}}$) in EGTA. Instead, we reuse the trajectories in the IID step to approximate $g_i^{\hat{i},-i}$ by ignoring the small changes in the state visitation density caused by $\pi_i \to \hat{\pi}_i$.

Remark 2. The meta-game $\mathcal{M}(\pi_i, \hat{\pi}_i, \pi_{-i}, \hat{\pi}_{-i})$ is a partially monotone game and has a pure strategy equilibrium, because the monotonic improvements $g_i^{i,-i} \leq g_i^{\hat{i},-i}$ and $g_{-i}^{i,-i} \leq g_{-i}^{\hat{i},-\hat{i}}$ when $\pi_i, \pi_{-i} \to \hat{\pi}_i, \hat{\pi}_{-i}$.

Taking the two-agent case as an example, as we can see in Eq. (5), meta-game $\mathcal{M}$ becomes a 2×2 matrix-form game, which is much smaller in size than the whole underlying game. Besides, according to Fig. 2 Right and Remark 2, all four cases have at least one pure strategy that leads a stable improvement direction. To this end, we can use the existing Nash solvers for matrix-form games to compute a Nash equilibrium $\rho_i, \rho_{-i} = \text{NashSolver}(\mathcal{M})$ for meta-game $\mathcal{M}$, where ρ_i and $\rho_{-i} \in [0,1]$, and the Nash equilibrium of the meta-game is also an approximated equilibrium of the restricted underlying game. Then, SIP policies $\bar{\pi}_i, \bar{\pi}_{-i}$ can be aggregated based on current policy π_i and predicted policy $\hat{\pi}_i$ in the IID for each agent i.

Assumption 1. *In the IID step, ILs enjoy the monotonic improvement against fixed opponent policies, in which the change from π_i to $\hat{\pi}_i$ is usually constrained by a small step size. Then, we assume that there is a linear, continuous and monotonic change in the restricted policy space between π_i and $\hat{\pi}_i$.*

---

[1] In this paper, we want to maximize the return, not minimize the loss, so we need to avoid a strict minimum.

In this case, with ρ_i being agent i's Nash equilibrium policy in the meta-game, $\bar{\pi}_i$ can be derived via a linear mixture: $\bar{\pi}_i = \rho_i \pi_i + (1 - \rho_i)\hat{\pi}_i$, which delimits agent i's SIP. Now, we can prove that $(\bar{\pi}_i, \bar{\pi}_{-i})$ is a weak stable fixed point for the underlying game in Theorem 1. Furthermore, based on Assumption 1, the payoff and policy space $[\pi_i, \hat{\pi}_i]$ for DRG are bounded in a linear continuous space, we can conclude the following theorem:

Theorem 1 (Existence of a Weak Stable Fixed Point). *If (ρ_i, ρ_{-i}) is a Nash equilibrium of the meta-game $\mathcal{M}$, then linear mixture joint policy $(\bar{\pi}_i, \bar{\pi}_{-i})$ is a weak stable fixed point for the DRG.*

According to Theorem 1, $(\bar{\pi}_i, \bar{\pi}_{-i})$ is a weak stable fixed point of the restricted underlying game. Although the weak stable fixed point is relatively weak compared to the stable fixed points [2], as we have stated, a weak stable fixed point is a reasonable (not as strong as it is rational) requirement for an algorithm to avoid the minimum. Furthermore, weak stable fixed points can suit general game settings. Similarly, a local Nash equilibrium can be stable or saddle in different games [20]. Therefore, the goodness of stable concepts depends on specific settings. If we make some additional game class assumptions, then we can easily obtain stronger fixed point types. Nevertheless, this approach comes with a cost, requiring additional computation or assumptions that may break the most general settings. In addition, when the meta-game has multiple Nash equilibria, an equilibrium is randomly selected in our work.

2.3   Improvement over a Weak Stable Fixed Point

Although the weak stable fixed point, $(\bar{\pi}_i, \bar{\pi}_{-i})$, binds the policy update to another fixed point, there are still fully stable points according to Theorem 1. Besides, it is difficult to generalize for the other parts of the policy space not reached by SIP, especially in anticoordination games. Similar to the extragradient method [21], to encourage the exploration, we apply the best response against the weak stable fixed point $(\bar{\pi}_i, \bar{\pi}_{-i})$:

$$\pi_i' = \arg\max_{\pi \in \Pi_{\theta_i}} \eta_i(\pi, \bar{\pi}_{-i}). \tag{6}$$

To perform the best response, we need another round to collect the experiences and perform a gradient step in Eq. (6). However, in practice, since we already have the trajectories in the IID step, the best response to the weak stable fixed point can be easily estimated through importance sampling. Alternatively, by defining $c_i \overset{\text{def}}{=} \min\left(1 + \bar{c}, \max(1 - \bar{c}, \frac{\pi_i(a_i|s)}{\bar{\pi}_i(a_i|s)})\right)$ as truncated importance sampling weights, we can rewrite the best response update to Eq. (6) as an equivalent form to the following one in terms of expectations: $\pi_i' = \arg\max_\pi \mathbb{E}_{a_{-i} \sim \bar{\pi}_{-i}}[c_{-i}\eta_i(\pi_i, \pi_{-i})]$. If the agents end up playing the BR, then there is no further improvement in the IID step; the payoff entries in the restricted meta-game would be zero, meaning agents will stay at the current policies following MATRL steps.

2.4 Local Convergence

MATRL is a gradient-based algorithm with the best response to policies within the SIPs, which is essentially a variant of LookAhead methods [42]. More specifically, MATRL enhances the classic LookAhead method with variable step size scaling or two time-scale update rules at each SIP step, which is controlled by restricted meta-game analysis. It has been proven that the LookAhead method can locally converge to a stable fixed point and avoid strict saddles in all differentiable games [15]. Similarly, we show the local convergence of MATRL in Theorem 2. Please note, here, that to investigate the convergence, fixed point iterations are conducted on the whole learning process, while the meta-game analysis step in MATRL borrows the variable stepsize scaling and shows it is reasonable to locally avoid unstable fixed points. Unlike LOLA, which uses a first-order Taylor expansion to estimate the best response to a predicted policy, we elaborately design the look-ahead step within the SIPs and perform the gradient steps for the best response to the SIPs. We also show that MATRL empirically outperforms the typical LookAhead method, IL LookAhead (IL-LA), in the experiments.

Theorem 2 (Local Convergence of MATRL). *Let the objectives $\eta_i(\pi_i, \pi_{-i})$ of agents are twice continuously differentiable and step size α is sufficiently small, MATRL converges locally to a stable fixed point with ϵ error in Euclidean distance.*

2.5 Discussions

Computation Cost. Compared to pure ILs, there are two extra cost sources in common meta-game analysis: approximating and solving the meta-game [23]. In our case, the meta-game is restricted to a local two-action game, where two actions, π_i and $\hat{\pi}_i$, are close to each other. Reusing the IID trajectories will some estimation errors [35], but this issue can be eased by large batch size. Then, we can enjoy this proximity property and reduce the meta-game approximation cost (without extra sampling) by reusing the collected trajectories in the IID step. The next crucial problem is how to solve the n-agent two-action meta-game, which consists of the 2^n entries of each of the n payoff matrices. Solving this meta-game is much simpler than solving the whole underlying game, which increases exponentially with state size, action size, agent number, and time horizons. As the general-sum matrix-form game has no fully polynomial time approximation for computing Nash equilibria, it usually costs a great deal to solve the game [6]. However, as shown in Remark 2, there always exists at least one pure Nash equilibrium in the meta-game, which can be computed in polynomial time [7]. Therefore, if we only require an approximated Nash equilibrium, then when n is small, for example, $n \leq 5$, it is affordable to find a meta-game Nash equilibrium with subexponential complexity.

Connections to Existing Methods. MATRL generalizes many existing methods with the best response. In extreme cases, where the meta-game Nash equilibrium is $(\rho_i, \rho_{-i}) = (1, 1)$, which means that the Nash aggregated policies always maintain the current policies, MATRL degenerates to ILs. Here, we always best respond to other agents' current policy π_i and $\pi'_i = \arg\max_{\pi_i} \eta_i(\pi_i, \pi_{-i})$ following Eq. (6). The LookAhead [8], extragradient [12] and exploitability descent [33] methods are also special instances of MATRL when meta-game Nash is $(\rho_i, \rho_{-i}) = (0, 0)$, which means that the best response to the most aggressive predicted policy $\hat{\pi}_{-i}$ and $\pi'_i = \arg\max_{\pi_i} \eta_i(\pi_i, \hat{\pi}_{-i})$. More specifically, let ξ denotes the game's simultaneous gradient, H_o is the matrix of anti-diagonal blocks of H (Hessian of the game), and α is step-size. Then we can have the updating gradient for LookAhead methods as $(I - \alpha H_o)\xi$. Similarly, for MATRL, we have the updating gradient $(I - \rho\alpha H_o)\xi$, where ρ is a ratio determined by meta-game Nash to dynamically adjust the step-size at each iteration.

3   Related Work

The research on the EGTA [35] creates a policy-space meta-game for modeling multi-agent interactions. Using various evaluation metrics, this work then updates and extends the policies based on the analysis of meta policies [14]. Although these methods are broad with respect to multi-agent tasks, they require extensive computing resources to estimate the empirical meta-game and solve it with its increasing size [24]. In our method, we adopt the idea of a policy-space meta-game to approximate the fixed point. Unlike previous works, we only maintain current and predicted policies to construct the meta-game, which is computationally achievable in most cases. The payoff entry in MATRL's meta-game is the expected advantage, which has a lower estimation variance compared to the commonly used empirically estimated return in EGTAs. Regardless, we can reuse the trajectories in the IID step to estimate the payoffs without incurring additional sampling costs.

Recently, due to the use of neural networks as a function approximation for policies and values, many works have emerged on deep reinforcement learning (DRL) [22]. TRPO [27,28] is one of the most successful DRL methods in the single-agent setting, which places constraints on the step size of policy updates, monotonically preserving any improvements. Based on the monotonic improvement in single-agent TRPO [27], MATRL extends the improvement guarantee to the multi-agent level towards a weak stable fixed point. Some works have directly applied fully decentralized single-agent DRL methods [32], which can be unstable during learning due to the issue of nonstationarity. [9,19,26,31] further exploit the setting of centralized learning decentralized execution (CTDE). These methods provide solutions for training agents in complex multi-agent environments, and the experimental results show their effectiveness compared with ILs. Similar to the CTDE setting, MATRL also enjoys fully decentralized execution. Although MATRL still needs knowledge about other agents' policies in adjusting the step size during training, it does not need centralized critics or any communication channels. Besides, [16] attempted to apply trust-region methods in networked multi-agent settings by conducting consensus optimization with their neighbors. Instead takes a game-theoretic approach to compute the meta-game Nash to find policy improvement directions without networked assumption.

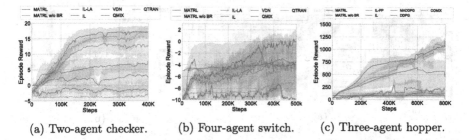

(a) Two-agent checker.   (b) Four-agent switch.   (c) Three-agent hopper.

Fig. 3. Learning curves in discrete and continuous tasks. The solid lines are average episode returns with 10 random seeds for each model, and the light color areas are the error bars.

4   Experiments

We design experiments to answer the following questions: 1) Can the MATRL method empirically contribute to convergence in general game settings, including cooperative/competitive and continuous/discrete games? 2) How is the performance of MATRL compared to ILs with the same hyperparameters and other strong MARL baselines in discrete and continuous games with various agent numbers? 3) Do the meta-game and best response to the weak stable fixed point bring about benefits? We first evaluate the convergence performance of MATRL in matrix form games to answer the first question and validate the effectiveness of convergence. For Question 2, we show that MATRL largely outperforms ILs (PPO [28]) and other centralized baselines (QMIX [26], QTRAN [30] and VDN [31]) in discrete grid world games that have coordination problems. MATRL also outperforms DDPG [17], MADDPG [19] and COMIX [39] for continuous multi-agent MuJoCo games. In addition, we test the algorithms with a 2-agent Atari Pong game to investigate whether MATRL can mitigate unstable cyclic behaviors [1] in zero-sum games. In these tasks, MATRL uses the same PPO configurations as ILs to examine the effectiveness of the trust region gradient-update mechanism, and we use official implementations for the other baselines. Finally, ablation studies are conducted by: 1. removing the best response, called the "MATRL w/o BR"; 2. skipping the SIP estimation, named "IL-LA", which has similar procedures as those of LOLA [8], which approximates the best response to the predicted policies via Taylor expansion, but IL-LA takes the best response gradient steps for the predicted policies. These configurations provide insights into how much, if at all, the SIP and the best response contribute to the MATRL's performance. The code and experiment scripts are also available at https://github.com/matrl-project/matrl.

Grid Worlds. We evaluated MATRL in two grid world games from MA-Gym [13], two-agent checker, and four-agent switch, which are similar to the games in [31] but with more agents to examine if MATRL can handle the games that have more than two agents. In the checker game, two agents cooperate in collecting fruit on the map; the sensitive agent obtains 5 for an apple and -5

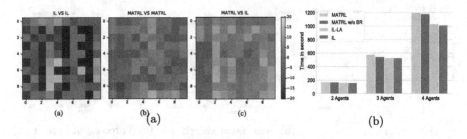

Fig. 4. (a): MATRL/IL versus MATRL/IL in the two-agent Pong game. For each setting, the grids show pairwise performance (average scores) by pitting their ten checkpoints against one another; yellow indicates a higher score. (b): Run time for 20,000 environment steps (including 50 gradient steps) for the algorithms in two- to four-agent games.

for a lemon, while the other agent obtains 1 and −1, respectively. Therefore, the optimal solution is to let the sensitive agent obtain the apple and the less sensitive agent obtain the lemon. In the four-agent switch game, two rooms are connected by a corridor, each room has two agents, and the four agents try to go through one corridor to the target in the opposite room. Only one agent can pass through the corridor at one time, and agents obtain −0.1 for each step and 5 for reaching the target, so they need to cooperate to obtain optimal scores. In both games, agents can move in four directions and only partially observe their position. Although our formulation uses a fully observable setting, in this game, the methods are adapted to the partially observable setting by pretending the observation is a state. We compare the MATRL with the PPO-based IL and two off-policy centralized training and decentralized execution baselines: VDN [31], QTRAN [30] and QMIX [26]. The results are given in Figs. 3a and 3b, where MATRL shows stable improvement and outperforms other baselines. In a two-agent checker game, using the best response, our method can achieve a total reward of 18, while the ILs' reward stays at −2. In addition, although PPO-based MATRL uses on-policy learning, it achieves better final results in fewer time steps compared to the off-policy baselines. For the four-agent switch game, as shown in Fig. 3b, MATRL can continuously improve the total rewards to 6.5, which is the closest to the optimal score for this game when compared with other baselines. The result of the four-agent switch also demonstrates the effectiveness of MATRL in guaranteeing stable policy improvement for games that have more than two agents.

Multi-agent MuJoCo. We also examined MATRL in a multi-agent continuous control task with a three-agent hopper from [39]. Here, three agents cooperatively control each part of a hopper to move forward. The agents are rewarded with the distance traveled and the number of steps they make before falling. Figure 3c shows that MATRL significantly outperforms ILs, MADDPG, DDPG, and the benchmarks like COMIX in [39] within the same amount of time.

Multi-agent Atari Pong Game. In the 2-agent Pong game experiments, we used raw pixels as observations and trained the MATRL and IL agents independently. Following training, we compare the pairwise performance of these models by pitting their ten checkpoints against one another and recording average scores. We report the results in Fig. 4a, which shows that MATRL outperforms ILs in MATRL vs. IL settings in most policy pairs. In addition, from the MATRL vs. MATRL and ILs vs. IL settings' results, we can see that MATRL has a more transitive learning process than that of ILs, which means that MATRL can mitigate the common cyclic behaviors in zero-sum games.

Effect and Cost of the SIP and Best Response to a Fixed Point. This section analyzes the effect of the SIP from the meta-game Nash equilibrium and the best response against the weak stable fixed point. The ablation settings are obtained by removing the SIP (IL-LA) and the best response (MATRL w/o BR). In Fig. 3, we can observe that in all the tasks, without the best response to the fixed point, the learning curves of MATRL w/o BR have higher variance and the lowest final scores. This establishes the importance of the best response to stabilize and improve agents' performance and empirically shows that MATRL has better convergence ability than do the other baselines. Additionally, without the SIP to select a fixed point, MATRL recovers to ILs with policy prediction (IL-LA) [8,42]. Similarly, the curves of IL-LA have lower final scores, and the convergence speed is not as good as that of MATRL, which suggests that the SIP provides benefits. MATRL w/o BR has lower variance compared to IL-LA, which reveals that the SIP can stabilize the learning via weak stable fixed point constraints. Finally, when compared to IL and IL-LA, as shown in Fig. 4b, in two- to four-agent games with 20,000 environment steps and 50 gradient steps, the training time of MATRL is empirically approximately 1.1–1.2 times slower. Given the significant performance improvement, we believe such extra computational cost from the SIP and the best response are acceptable.

5   Conclusions

We proposed and analyzed the trust region method for multi-agent learning problems, which considers the IID and SIP to meet multi-agent learning objectives. In practice, based on independent trust payoff learners, we provide a convenient way to approximate a further restricted step size within the SIP via a meta-game. This approach ensures that MATRL is generalized, flexible, and easily implemented to deal with multi-agent learning problems in general. Our experimental results justify the fact that the MATRL method significantly outperforms ILs using the same configurations and other strong MARL baselines in both continuous and discrete games with varying numbers of agents.

References

1. Balduzzi, D., Garnelo, M., Bachrach, Y., Czarnecki, W., et al.: Open-ended learning in symmetric zero-sum games. In: ICML, vol. 97, pp. 434–443. PMLR (2019)
2. Balduzzi, D., Racanière, S., Martens, J., Foerster, J.N., et al.: The mechanics of n-player differentiable games. In: ICML, vol. 80, pp. 363–372. PMLR (2018)

3. Berner, C., Brockman, G., Chan, B., Cheung, V., et al.: Dota 2 with large scale deep reinforcement learning. arXiv preprint arXiv:1912.06680 (2019)
4. Bowling, M., Veloso, M.: Existence of multiagent equilibria with limited agents. J. Artif. Intell. Res. **22**, 353–384 (2004)
5. Buşoniu, L., Babuška, R., De Schutter, B.: Multi-agent reinforcement learning: an overview. In: Srinivasan, D., Jain, L.C. (eds.) Innovations in Multi-agent Systems and Applications - 1, pp. 183–221. Springer, Heidelberg (2010). https://doi.org/10.1007/978-3-642-14435-6_7
6. Daskalakis, C., Goldberg, P.W., Papadimitriou, C.H.: The complexity of computing a nash equilibrium. SIAM J. Comput. **39**(1), 195–259 (2009)
7. Fabrikant, A., Papadimitriou, C., Talwar, K.: The complexity of pure nash equilibria. In: STOC, pp. 604–612 (2004)
8. Foerster, J., Chen, R.Y., Al-Shedivat, M., Whiteson, S., et al.: Learning with opponent-learning awareness. In: AAMAS, pp. 122–130 (2018)
9. Foerster, J.N., Farquhar, G., Afouras, T., Nardelli, N., et al.: Counterfactual multi-agent policy gradients. In: AAAI, pp. 2974–2982. AAAI Press (2018)
10. Fudenberg, D., Drew, F., Levine, D.K., Levine, D.K.: The Theory of Learning in Games, vol. 2. MIT Press (1998)
11. Hernandez-Leal, P., Kaisers, M., Baarslag, T., de Cote, E.M.: A survey of learning in multiagent environments: dealing with non-stationarity. arXiv preprint arXiv:1707.09183 (2017)
12. Kakade, S.M., Langford, J.: Approximately optimal approximate reinforcement learning. In: ICML, pp. 267–274. Morgan Kaufmann (2002)
13. Koul, A.: A collection of multi agent environments based on OpenAI gym (2019). https://github.com/koulanurag/ma-gym.git
14. Lanctot, M., Zambaldi, V.F., Gruslys, A., Lazaridou, A., et al.: A unified game-theoretic approach to multiagent reinforcement learning. In: NeurIPS, pp. 4190–4203 (2017)
15. Letcher, A., Foerster, J.N., Balduzzi, D., Rocktäschel, T., et al.: Stable opponent shaping in differentiable games. In: ICLR (2019)
16. Li, W., Wang, X., Jin, B., Sheng, J., et al.: Dealing with non-stationarity in multi-agent reinforcement learning via trust region decomposition. arXiv preprint arXiv:2102.10616 (2021)
17. Lillicrap, T.P., Hunt, J.J., Pritzel, A., Heess, N., et al.: Continuous control with deep reinforcement learning. In: ICLR (2016)
18. Littman, M.L.: Markov games as a framework for multi-agent reinforcement learning. In: Machine Learning Proceedings, pp. 157–163. Elsevier (1994)
19. Lowe, R., Wu, Y., Tamar, A., Harb, J., et al.: Multi-agent actor-critic for mixed cooperative-competitive environments. In: NeurIPS, pp. 6379–6390 (2017)
20. Mazumdar, E., Ratliff, L.J., Sastry, S.S.: On gradient-based learning in continuous games. SIAM J. Math. Data Sci. **2**(1), 103–131 (2020)
21. Mertikopoulos, P., Lecouat, B., Zenati, H., Foo, C., et al.: Optimistic mirror descent in saddle-point problems: going the extra (gradient) mile. In: ICLR (2019)
22. Mnih, V., Kavukcuoglu, K., Silver, D., Graves, A., et al.: Playing atari with deep reinforcement learning. arXiv preprint arXiv:1312.5602 (2013)
23. Muller, P., Omidshafiei, S., Rowland, M., Tuyls, K., et al.: A generalized training approach for multiagent learning. In: ICLR (2020)
24. Omidshafiei, S., Papadimitriou, C., Piliouras, G., Tuyls, K., et al.: α-rank: multi-agent evaluation by evolution. Sci. Rep. **9**(1), 1–29 (2019)

25. Peng, P., Wen, Y., Yang, Y., Yuan, Q., et al.: Multiagent bidirectionally-coordinated nets: emergence of human-level coordination in learning to play starcraft combat games. arXiv preprint arXiv:1703.10069 (2017)
26. Rashid, T., Samvelyan, M., de Witt, C.S., Farquhar, G., et al.: QMIX: monotonic value function factorisation for deep multi-agent reinforcement learning. In: ICML, vol. 80, pp. 4292–4301 (2018)
27. Schulman, J., Levine, S., Abbeel, P., Jordan, M.I., et al.: Trust region policy optimization. In: ICML. JMLR Workshop and Conference Proceedings, vol. 37, pp. 1889–1897. JMLR.org (2015)
28. Schulman, J., Wolski, F., Dhariwal, P., Radford, A., et al.: Proximal policy optimization algorithms. CoRR abs/1707.06347 (2017)
29. Shoham, Y., Leyton-Brown, K.: Multiagent Systems: Algorithmic, Game-Theoretic, and Logical Foundations. Cambridge University Press (2008)
30. Son, K., Kim, D., Kang, W.J., Hostallero, D., et al.: QTRAN: learning to factorize with transformation for cooperative multi-agent reinforcement learning. In: ICML, vol. 97, pp. 5887–5896. PMLR (2019)
31. Sunehag, P., Lever, G., Gruslys, A., Czarnecki, W.M., et al.: Value-decomposition networks for cooperative multi-agent learning based on team reward. In: AAMAS, pp. 2085–2087 (2018)
32. Tan, M.: Multi-agent reinforcement learning: Independent vs. cooperative agents. In: AAAI, pp. 330–337 (1993)
33. Tang, J., Paster, K., Abbeel, P.: Equilibrium finding via asymmetric self-play reinforcement learning. In: Deep Reinforcement Learning Workshop NeurIPS 2018 (2018)
34. Terry, J.K., Black, B.: Multiplayer support for the arcade learning environment. arXiv preprint arXiv:2009.09341 (2020)
35. Tuyls, K., Perolat, J., Lanctot, M., Hughes, E., et al.: Bounds and dynamics for empirical game theoretic analysis. Auton. Agent. Multi-agent Syst. **34**(1), 7 (2020)
36. Tuyls, K., Perolat, J., Lanctot, M., Leibo, J.Z., et al.: A generalised method for empirical game theoretic analysis. In: AAMAS, pp. 77–85 (2018)
37. Vinyals, O., Babuschkin, I., Czarnecki, W.M., Mathieu, M., et al.: Grandmaster level in starcraft II using multi-agent reinforcement learning. Nature **575**(7782), 350–354 (2019)
38. Wellman, M.P.: Methods for empirical game-theoretic analysis. In: AAAI, pp. 1552–1556 (2006)
39. de Witt, C.S., Peng, B., Kamienny, P.A., Torr, P., et al.: Deep multi-agent reinforcement learning for decentralized continuous cooperative control (2020)
40. Yang, Y., Wang, J.: An overview of multi-agent reinforcement learning from game theoretical perspective. arXiv preprint arXiv:2011.00583 (2020)
41. Yang, Y., Wen, Y., Wang, J., Chen, L., et al.: Multi-agent determinantal q-learning. In: ICML 2020, pp. 10757–10766. PMLR (2020)
42. Zhang, C., Lesser, V.R.: Multi-agent learning with policy prediction. In: AAAI. AAAI Press (2010)
43. Zhou, M., Luo, J., Villela, J., Yang, Y., et al.: Smarts: scalable multi-agent reinforcement learning training school for autonomous driving. arXiv preprint arXiv:2010.09776 (2020)

An Adaptive Negotiation Dialogue Agent with Efficient Detection and Optimal Response

Qisong Sun and Siqi Chen[(✉)]

College of Intelligence and Computing, Tianjin University, Tianjin, China
siqichen@tju.edu.cn

Abstract. For negotiation dialogue tasks, instead of adopting stationary strategies, a more advanced opponent may demonstrate sophisticated behaviors by employing reasoning strategies to predict its opponent's actions. To address this challenge, this work proposes a novel dialogue agent, which leverages the predictive power of Bayesian policy reuse and the recursive reasoning ability of theory of mind, allowing efficiently detecting the policy of opponents using either stationary or higher-level reasoning strategies and learning a best-response policy when faced with previously unseen strategies. Finally, we present the results of the proposed agent against state-of-the-art baselines on the CRAIGSLISTBARGAIN dataset and show that the agent outperforms existing agents and its efficacy of detecting new unseen strategies (This is an extended version of the paper [3] presented at the 20th IEEE International Conference on Ubiquitous Intelligence and Computing 2022.)

Keywords: Reinforcement Learning · Dialogue system · Negotiation agent · Multi-agent system · Bayesian Policy Reuse · Theory of mind

1 Introduction

The task of building dialogue systems for negotiation is challenging as it demands both effective communication and strategic reasoning [12]. Despite the success of deep learning [6,19] and reinforcement learning in generating useful dialogue strategies [2,9,14], most work assumes that opponents employ stationary strategies rather than more complex strategies that adjust behaviors based on the opponent. This assumption, however, limits their applicability to realistic scenarios where opponents may reason about each other's strategies and react optimally, e.g., during negotiation, one side can infer the intention of the opponent and predict the effect of their own words on the opponent's state and behavior.

In response to these limitations, this study presents a dialogue agent called the BayesToM agent, based on the Bayesian Theory of Mind (BayesToM) algorithm on Policy. [24]. Specifically, our BayesToM agent combines the Bayesian policy reuse (BPR) [18] and theory of mind (ToM) [20]. ToM is a recursive reasoning technique that describes a cognitive mechanism for explicitly attributing unobservable mental contents to other players, such as beliefs and intentions, to help predict their actions. BPR is used for responding to a new opponent by selecting a policy from its policy library, and maintaining a probability distribution (i.e., Bayesian belief) over a set of known opponent strategies capturing their similarity to the new policy, which is updated with observed signals related with the performance of a policy. Unlike vanilla BPR, which can only detect non-stationary opponents, the BayesToM agent can quickly and accurately detect non-stationary and more sophisticated opponents that may construct an opponent model, and optimizes its own behavior by using the predictive power of BPR and the recursive inference ability of ToM.

In the face of unseen strategies, the BayesToM agent demonstrates its ability to detect and learn optimal response strategies. Evaluations were conducted on the CRAIGSLISTBARGAIN dataset [9]. The results indicate that the superiority of BayesToM over other dialogue agents in terms of completion rates and utility achieved. Additionally, the experiments confirm the BayesToM agent's proficiency in detecting and adapting to an opponent's new strategy.

The rest of this work is organized as follows. Section 2 overviews important related work. Section 3 presents the framework of the negotiation dialogue, which is composed of the task description, MDP formulation and negotiation systems. Subsequently, the details of our agent are given in Sect. 4. In Sect. 5, extensive experimental results are reported to verify the effectiveness of our agent. Finally, Sect. 6 concludes and discusses future research directions.

2 Related Work

2.1 Negotiation Dialogues

Negotiation settings in which agents use natural language to bargain have attracted much research attentions. There are a large body of work covering various aspects of negotiation [2,4,5,22]. Lewis et al. [14] trained negotiation agents based on an end-to-end RNN model to divide a set of items, e.g., books, hats and basketballs. This study is grounded in a closed-domain game with a fixed set of objects, thus lacking richness of human behavior. Moreover, the agent's goal is directly optimized through reinforcement learning (RL), which often results in degenerate solutions where the utterances become ungrammatical. To study human negotiation in more open-ended settings that involve real goods, He et al. [9] collected a new dataset (CRAIGSLISTBARGAIN) of negotiation dialogues by using Amazon Mechanical Turk for a task where two workers were assigned the roles of buyer and seller respectively, and bargained the price of an items for sale on craigslist.org. Their negotiation agent based on RL was reported to be more effective than other competitors. Yang et al. [23] further improved negotiation agents' performance by modeling opponent personality. A Theory of Mind based model was applied to predict an opponent's response given the current state, allowing one-step lookaheads during inference. The ToM agent achieved a higher dialogue agreement rate and utility compared to the used benchmarks. However, it only considers the opponent's primitive actions instead of high-level strategies, which leads to slow adaptation to opponents that are non-stationary.

2.2   Bayesian Policy Reuse

Bayesian Policy Reuse (BPR) was first presented in [18] as a framework for an agent to swiftly choose the optimal policy to execute when faced with an unknown opponent based on the opponent's policy. A model-based algorithm termed BPR+ in repeated games is later refined by [11] with the potential to expand the policy library online. BPR+ has the limitation of only using episodic rewards as signals, which makes it difficult to determine the opponent type in challenging tasks. With integration of BPR and Pepper [7], Bayes-Pepper [10] has been proposed for use in Markov games. When the opponent's policy changes during an episode, Bayes-Pepper employs state-action pairs as additional signals to modify its policy. BPR+ and Bayes-Pepper may be ineffective when addressing complicated scenarios because they are tabular-based algorithms. Zheng et al. [26]] proposed Deep BPR+ by extending BPR+ [11] with DRL techniques [16]. With the help of opponent modeling, Deep BPR+ significantly increases detection accuracy by using a policy distillation network as the policy library. However, Deep BPR+ exclusively concentrates on opponents who switch randomly between a range of strategies, which is a major limitation of the method, e.g., after a number of interactions, the opponent may intelligently change its policy to seek higher returns.

2.3   Theory of Mind

The theory of mind (ToM) [20] can help predict an opponent's action, by constructing an abstract model of the opponent using recursive nested beliefs. ToM has been studied for training RL-based dialogue systems and to make dialogue systems interpretable [1]. More recently, the theory of mind has been applied to learn dialogue policy in certain domains. For navigation situations where questions and answers are created between a traveling agent and a guiding agent, the Recursive Mental Model [17] was presented. Another approach, called Answer in Questioner's Mind [13], used information-theoretic techniques to tackle a response guessing game.

A zero-order ToM (i.e., ToM_0) agent is unable to model the mental content of its opponents, and holds the zero-order belief in the form of a probability distribution over the action set of its opponent. The ToM_0 agent simply maximizes its expected payoff depending on the zero-order belief. In contrast, a first-order theory of mind (ToM_1) agent considers the possibility that its opponent is trying to win the game in a more disciplined way, and it reacts to the choices given by the ToM_1 agent in the previous interactions. A ToM_1 agent thus keeps both zero-order belief and first-order belief (a probability distribution that describes what the ToM_1 agent believes its opponents believes about itself), and combines the two beliefs based on first-order confidence to predict opponents' behaviors, where the first-order confidence is adjusted based on the results.

3 Framework

3.1 Task

This work considers a task in which there is a bilateral negotiation dialogue between a buyer and a seller. The task is based on the dataset CRAIGSLIST-BARGAIN [9], which consists of 6682 human-human negotiation dialogues scraped from craigslist.org about a buyer and a seller bargaining the price of an item, including information like product description, photos, and the listing price. Both sides are encouraged to get the best deal for itself (in terms of utility). This dataset covers realistic scenarios with rich and diverse negotiation behavior, e.g., cheap talks, embellishment and side offers. The item's description is given to both parties, whereas each party's target price is kept private. Negotiation process continues until an agreement is reached or one side breaks off.

Table 1. Dialogue acts based on the CRAIGSLISTBARGIN dataset.

Dialog Act	Definition	Example
greet	say hello or chat randomly	Hello, are you interested in the bed?
inquire	ask questions about the product, such as the usage, quality, parameters, etc	How about the cushions? Are they in good shape? Do they have any smell?
inform	answer questions about products	It was only used a few times. it is in excellent condition
propose(price=)	initiate a price or a price range for the product	Oh, I've own a piece from them in the past, it only lasted 5 years. Would you take $30?
counter(price=)	propose a new price or a new price range	That's very low for me. I got the item for $80. The couch is very comfortable and in a very good condition. The item has very strong woods that last for years
counter-noprice	dissatisfied with the current price, but does not specify the desired price	They're nice but expensive. I prefer this machine to paying them if you know whta I mean
confirm	confirmation of certain aspects of products	The bathroom is pained white, and the combo shower/bath is also white
affirm	give an affirmative response to a confirm/propose	Sweet. Is there anyway you'd allow me to bring my dog? I swear she never pees in the house
deny	give a negative response to a confirm/propose	No scratches in excellent shape, I do have a matching night table i can include if interested
agree(price=)	make a deal	That's a good deal. Can you lower the price to 180 and i pick up and it's a deal
disagree(price)	cannot make a deal	Understood. The asking price is $24, but if you could give me $15 we could make a deal no
offer(price=)	final offer with price, no utterance	OFFER($1500)
accept	final acceptance, no utterance	ACCEPT
reject	final rejection, no utterance	REJECT
quit	leave the negotiation, no utterance	QUIT

3.2 Negotiation Systems

As shown in Fig. 1, our negotiation system consists of three important modules, which is in line with traditional goal-oriented dialogue systems [25].

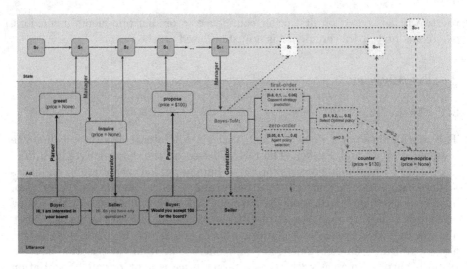

Fig. 1. The framework of negotiation systems.

- A **parser** resolves the utterance sent by the opponent and outputs the high level dialogue act, which serves as a module for natural language understanding. Dialogue acts are defined by an intent and its arguments. For example, it converts the utterance "I really can't go that low. I just renovated the kitchen and our tenants are great people." to the dialog act – counter-noprice. Note that dialogue acts in this work are intended to capture the high level strategic moves taken during the negotiation process, rather than the complete semantics of a sentence. Table 1 overviews the dialogue acts.
- A **manager** is the core of the whole system, which governs the negotiation moves and responds to the opponent's utterances with the dialogue act (e.g., proposing a new offer, accepting a counter offer, etc.). Our Bayes-ToM approach is applied to the dialog manger.
- A **generator** servers a role as natural language generation which composes the dialogue act given by the manger into next utterance to give.

In our implementation, we employ Microsoft Language Understanding Intelligent Service (LUIS) [21] with 10 annotated training samples for each dialog act as the utterance parser. The generator uses a retrieval-based method condition on both the current dialogue act suggested by the manger and the dialogue history, e.g., counter(price = $810) + dialogue history → How about $810 for the rent?. To promote linguistic diversity, it samples an utterance from the top 10 matched candidates.

The basic type of the dialog manger is the one trained using supervised learning (SL) [9]. The SL manger uses a neural network to represent state transitions and employs vanilla behavior cloning to learn a negotiation strategy directly from the negotiation corpus with hand-coded rules to fine-tune the learned strategy (e.g., filling prices in the arguments of an intent). It ensures that a proposal will be accepted if the current price is at least 70% of the target price. An advanced one is the RL manger further refines the SL model using reinforcement learning algorithm called Advantage Actor-Critic [15]. Its policy network is initialised by the SL manger without applying any rules. The RL manger considers neither self-play to enhance strategy diversity nor value distribution to enhance policy evaluation. The third one is the ToM manager [23], which is based on the RL model and uses a probabilistic state representation to learn a response policy that takes an opponent's personality and mental state into account.

Next, we will introduce the details of our approach, which can efficiently detect the strategy of opponents using either stationary or higher-level reasoning strategies.

4   The Approach

The overview of our dialog agent is given in Algorithm 1. In this section, we concentrate on the core part of framework—the manger—decides on the responding dialogue act given the current state. We start with the Markov Decision Process formulation, which defines our task.

4.1   MDP Formulation

We formulate our negotiation process as a Markov Decision Process (MDP), $< \mathcal{N}, S, A, T, R >$, with $\mathcal{N} = \{-1, 1\}$ denoting the set of two agents (i.e., buyer $= -1$/seller $= 1$), a set of states S, a set of dialogue acts A, a deterministic transition function T, a deterministic reward function R. S contains all negotiation states, and each $s_t \in S$ encodes the latest dialogue act and the preceding acts, e.g., $s_t = (s_0, a_1^i, a_2^{-i}, ..., a_{t-1}^i, a_t^{-i})$. A contains all dialogue acts, and each $a_i \in A$ encodes the *intent* (inform, counter, offer, etc.) of an utterance and the *price* (if any). For example, OFFER($\$3200$) expresses that the agent proposes a formal offer with price $\$3200$. T formulates the state transition based on the current state and the intended dialogue act.

4.2   Bayesian Theory of Mind for Dialogue

When incorporating theory of mind with Bayesian policy reuse, the simplest one is the zero-order case – Bayes-ToM$_0$, which is essentially equivalent to the vanilla BPR with new Opponent detection and learning. Specifically, Bayes-ToM$_0$ maintains a zero-order belief β^0 about opponent's strategies, where β^0 is a probability distribution over previously seen strategies. To update β^0, Bayes-ToM$_0$ uses a performance model $P(U|j, \pi)$ to describe the performance of policy π, i.e., a probability distribution over the return U using π on opponent strategy j. The approach selects the best responding policy through BPR-EI heuristics, and updates its zero-order belief using received reward signals. When an unknown opponent strategy is detected, Bayes-ToM$_0$ switches to learn a new policy against it. Due to space constraints, more details about the BPR method can be found in the previous work [22].

Now, we turn attention to the first-order case (Bayes-ToM$_1$)[1]. Since complicated opponents may predict our policy and then make best responses towards the estimated policy accordingly, Bayes-ToM$_1$ holds a first-order belief model β^1 in addition to a zero-order belief β^0 to cope with such complicated opponents. The first-order belief model β^1 captures the probability that our agent believes the opponent believes it will choose a policy $\pi \in \Pi$, which is updated on the basis of the performance mode $P_{oppo}(U|J, \Pi)$ (corres). The Bayes-ToM$_1$ agent first predicts its opponent's strategy $\hat{j}$ under the assumption that the opponent maximizes its own utility based on the BPR-EI heuristic with its first-order belief (corresponding to line 4 of Algorithm 1). This prediction's outcome, nevertheless, might be inconsistent with what its zero-order belief predicts. Therefore to balance the effect between the agent's first-order prediction and the zero-order belief, a first-order confidence c_1 ($0 \le c_1 \le 1$) is employed. The final prediction is then calculated using a linear weighted combination of the first-order prediction $\hat{j}$ and the zero-order belief $\beta^{(0)}$, controlled by the confidence parameter c_1 following Eq. 1 given in [20] (corresponding to line 4).

$$I(\beta^{(0)}, \hat{j}, c_1)(j) = \begin{cases} (1 - c_1)\beta^{(0)}(j) + c_1 & \text{if } j = \hat{j} \\ (1 - c_1)\beta^{(0)}(j) & \text{otherwise} \end{cases} \tag{1}$$

With this integrated belief I, the best response policy can be chosen (corresponding to line 5). Then, the first-order belief and zero-order belief are both updated following Bayes' rule [18] (corresponding to line 8–15).

In the previous ToM model [20], the value of c_1 would rise if the forecast was accurate, and vice versa. However, this information is not available in our settings since agents are reluctant to reveal their policies to others to avoid exploitation in competitive environments like negotiations. Under first-order belief c_1 can be interpreted as the exploration rate of the opponent's strategy, we thus use negotiation outcomes as an indicator of whether the previous predictions are correct. To be specific, the value of c_1 increases when the agent wins (i.e., $r_{self} > r_{oppo}$), and it decreases by an amount of λ, otherwise. This heuristic works well when playing against Bayes-ToM$_0$, as the theory of mind technique can provide Bayes-ToM$_1$ with the optimal policy. However, for those opponents who randomly switch among several fixed policies or are incapable of using ToM, it may be ineffective because of oscillation of the curve of c_1.

Therefore, we use a generalized way to adjust c_1 using the concept of win rate $w_i = \frac{\sum_{i-l}^{i} r_{self}}{l}$. The value of l controls the number of episodes before considering that our Bayes-ToM$_1$ agent's opponent has a less sophisticated policy, since it starts out by assuming that its opponent is a Bayes-ToM$_0$ agent. We increase the value of c_1 by the adjustment rate λ, if the average performance until the current episode is better than the previous episode ($w_i \ge w_{i-1}$). When w_i is less than w_{i-1} but still above the threshold δ, this indicates a decrease in the performance of first-order predictions. Bayes-ToM$_1$ then rapidly decreases the

---

[1] Please note that as [20] proved that the reasoning levels that are deeper than level 2 cannot provide further benefits, here we focus on Bayes-ToM$_1$.

Algorithm 1. Bayes-ToM$_1$ Negotiation Agent

Parser: Converts the opponent's utterance u_{t-1}^{-i} to dialogue act a_{t-1}^{-i} and transmit it to the manager

Manager:

1: Initialize policy library Π and opponent strategies J, performance mode $P_{self}(U|J,\Pi)$ and $P_{oppo}(U|J,\Pi)$, zero-order belief $\beta^{(0)}$, first-order belief $\beta^{(1)}$

2: **for** each episode **do**

3:    Receive the dialog act a_{t-1}^{-i} from the parser

4:    Compute the first-order opponent policy prediction $\hat{j}$:

5:    $\arg max_{j \in J} \int_{\underline{u}}^{u^{max}} \sum_{\pi \in \Pi} \beta^{(1)}(\pi) P_{oppo}\left(U^+|\pi,\ j\right) dU^+$

6:    Integrate $\hat{j}$ with $\beta^{(0)}$: $I(\beta^{(0)}, \hat{j}, \mu)$

7:    Select the optimal policy π^*:

8:    $\arg max_{\pi \in \Pi} \int_{\underline{u}}^{u^{max}} \sum_{j' \in J} I(\beta^{(0)}, \hat{j}, \mu)(j) P_{self}\left(U^+|j,\ \pi\right) dU^+$

9:    Play and receive the episodic return $\langle r_{self}, r_{oppo}\rangle$

10:    **for** each own policy $\pi \in \Pi$ **do**

11:       Update first-order belief $\beta^{(1)}(\pi)$:

12:       $\beta^{(1)}(\pi) = \frac{P_{oppo}(r_{oppo}|\pi,\hat{j})\beta^{(1)}(\pi)}{\sum_{\pi' \in \Pi} P_{oppo}(r_{oppo}|\pi',\hat{j})\beta^{(1)}(\pi')}$

13:    **end for**

14:    **for** each opponent policy $j \in J$ **do**

15:       Update zero-order belief $\beta^{(0)}(j)$:

16:       $\beta^{(0)}(j) = \frac{P_{self}(r_{self}|j,\pi)\beta^{(0)}(j)}{\sum_{j' \in J} P_{self}(r_{self}|j',\pi)\beta^{(0)}(j')}$

17:    **end for**

18:    Update c_1

19:    Detect new opponent policy

20:    Generate dialog act a_t^i based on first-order theory and zero-order theory

21: **end for**

Generator: Produces natural language response u_t^i based on the current dialogue act a_t^i and the dialog state s_{t-1}

---

value of c_1 by a decreasing factor $\frac{\lg w_i}{\lg (w_i-\delta)}$; if $w_i \leq \delta$, we set the rate of exploring first-order beliefs to 0 and only utilize zero-order beliefs for prediction since we believe that the opponent uses only simple strategies and does not model our policy. So, we have c_1 as follows,

$$c_1 = \begin{cases} ((1-\lambda)\mu + \lambda)\mathbf{P}(w_i) & \text{if } w_i \geq w_{i-1} \\ (\frac{\lg w_i}{\lg (w_i-\delta)}\mu)\mathbf{P}(w_i) & \text{if } \delta < w_i \leq w_{i-1} \\ \lambda \mathbf{P}(w_i) & \text{if } w_i \leq \delta \end{cases} \quad (2)$$

where δ is a win rate threshold that represents a lower bound on the difference between an agent's forecast and its opponent's actual actions. The direction of the c_1 value's adjustment is controlled by the indicator function $\mathbf{P}(w_i)$. Assuming that its win rate v_i is below δ, Bayes-ToM$_1$ recognizes the change in the opponent's policy in each event I and reverses the value of $\mathbf{P}(w_i)$. Finally, Bayes-ToM$_1$

learns a new optimal policy if it detects a new opponent policy (detailed in next section).

In fact, the Bayes-ToM$_1$ agent is identical to the Bayes-ToM$_0$ agent in the new opponent detection and learning component. After realizing the opponent is employing a new policy, the agent turns to the learning stage and starts to learn the best-response policy against it. The Bayes-ToM agent considers Soft Actor Critic (SAC) algorithm [8] to do off-policy learning using the obtained interaction experience. The objective of SAC is to maximize the expected return and the entropy at the same time:

$$J(\theta) = \sum_{t=1}^{T} \mathbb{E}_{(s_t,a_t)\sim\rho_{\pi_\theta}} [r(s_t, a_t) + \alpha E(\pi_\theta(.|s_t))] \tag{3}$$

where α is the temperature parameter which controls how important the entropy term is and E(.) is the entropy measure. Entropy is a measure of the stochastic in the policy and helps to improve the robustness and universality of the trained model. The policy is trained to maximize the trade-off between expected return and entropy. After obtaining the learned policy, the agent updates its policy libraries Π and its opponent's policy library $\mathcal{J}$, respectively.

The Bayes-ToM agent determines whether its opponent is employing a new policy by recording negotiation results of fixed length. The number of episodes to be considered before deciding if an opponent employs a new undiscovered policy is represented by h. The Bayes-ToM agent records negotiation outcomes h in episode i and utilizes the win rate $\theta_i = \frac{\sum_{i-h}^{i} r^{self}}{h}$ from the most recent h episodes as a signal of the average performance of all policies up to the present i episode. If the win rate θ_i is below a specified threshold δ ($\theta_i < \delta$), the Bayes-ToM agent concludes that the opponent is employing a previously unseen policy.

5   Experiments

5.1   Experimental Setup

Opponents. Four types of opponents are considered in our experiments as introduced in Sect. 3.2, 1) an TOM(e) agent that is equipped with the explicit first-order ToM manager, where the opponent's personality can be estimated to a particular type from dialogs, 2) an ToM(i) agent that implements the implicit first-order ToM manager, where the opponent type is modelled as a latent variable, 3) an agent with SL manager, which uses a LSTM network to learn the transitions from s_t to s_{t-1} and 4) an agent with RL manager which employs an actor-critic method [15] to fine-tune its behavior.

Training. We trained our Bayes-ToM agent with a learning count of 10,000 times for each training, and seven different SL-based opponents with different rules (e.g., representing behaviors from cooperative to competitive types) for changing prices and rendering utterances are used in the training following the settings as in [23]. Since DQN is a classic deep reinforcement learning and is used in other work related to negotiation, we also implement another version of Bayes-ToM agent that is equipped with DQN to compare the effect of SAC algorithm. The maximum number of dialogues is set at 20, as most dialogues end within 10 rounds. If no agreement is reached after 20 rounds, the negotiation ends with failure and both sides of the negotiation will be equally penalized.

Both the seller's and the buyer's rewards are linear functions of the agreed price, where the seller's reward is 1 if the seller achieves its target price (i.e., the listing price of the item) and the buyer's reward is 1 if the buyer gets its target price (i.e., the lowest price it expects). The rewards for both sides are zero when the agreed price is the middle of the two prices above. The entire process can be seen as a zero-sum game. If no agreement is achieved, both the buyer and the seller suffer the same penalty of -0.5.

Evaluation Metric. We evaluate the outcomes of negotiation dialogues across four aspects:

1. Agreement rate (Ag): is defined as the ratio between successful negotiations and all negotiations.
2. Objective utility (Ut): is given by

$$\text{Ut}^i = \begin{cases} (P_{deal} - P_{target}^{-i})/\Delta P & \text{deal;} \\ 0 & \text{no deal} \end{cases}$$

   where P_{deal} stands for the the agreed price, P_{deal}^{-i} for the opponent's target price, and ΔP the difference between the target prices of the agent and its opponent.
3. Deal fairness (Fa): measures the fairness of an outcome, $\text{Fa}^i = 1 - 2*|\text{Ut}^i - 0.5|$
4. Dialog length (Len): represents the length of a dialog in terms of rounds.

5.2 Performance Against Different Chatbots

Results Against State-of-the-Art Dialogue Agents. To validate the effectiveness of Bayes-ToM$_1$, Bayes-ToM$_1$ agent is first compared with four opponents of various policies as introduced in Sect. 5.1. Moreover, the experiment also considers Bayes-ToM$_1$ with DQN.

The results are depicted in Fig. 2. As depicted in the figure, Bayes-ToM$_1$ with SAC demonstrated excellent overall performance against all the opponents, which was the best performing agent from the perspective of agreement rat, utility and fairness. More precisely, this agent achieved an average 27% improvement in Ag compared to other agents and a 43% improvement in Ut compared to the averaged score of other agents. The detection ability of Bayesian policy reuse and the inference power of the theory of mind may account for this success. The performance of Bayes-ToM$_1$ with SAC algorithm was better than that of

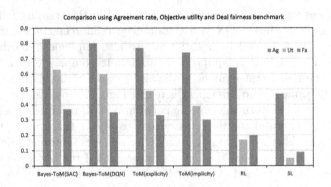

Fig. 2. The performance of our agent against other complex agents

Bayes-ToM$_1$ with DQN algorithm, due to the fact that the strategies learned by SAC algorithm have certain randomness compared to DQN, which can achieve higher returns in some scenarios. For both ToM agents, the employment of theory of mind aids them to model their opponents and thus adjust their strategies to obtain higher Ut and Ag, only being second to Bayes-ToM$_1$. The SL agent performed the worst among all agents, with the lowest Ag, Ut and Fa, mainly because the SL agent was trained to clone human behaviors, which failed to generate appropriate strategies.

Results Against Advanced Dialogue Agents. In order to study the advantage of Bayes-ToM$_1$ over Bayes-ToM$_0$ when play against advanced agents that can model opponents, ToM$_e$, ToM$_i$ and Bayes-ToM$_0$ were selected as negotiation opponents, and the results are shown in Table 2.

Table 2. Negotiation results for Bayes-ToM with ToM (implicit), ToM (explicit) and Bayes-ToM$_0$

agents	ToM (implicit)				ToM (explicit)				Bayes-ToM$_0$			
	Ag	Ut	Fa	Len	Ag	Ut	Fa	Len	Ag	Ut	Fa	Len
Bayes-ToM$_0$	0.8	0.24	0.48	10.7	0.8	0.27	0.54	9.4	0.6	0.1	0.2	10.2
Bayes-ToM$_1$	0.83	0.59	0.82	9.25	0.89	0.62	0.76	10.6	0.88	0.68	0.64	8.9

As can be seen from the table, Bayes-ToM$_1$ had a better overall performance than Bayes-ToM$_0$ when negotiating with both ToM(i) and ToM(e). Especially, Bayes-ToM$_1$ led by a significant margin in Ut and Fa. When facing Bayes-ToM$_0$, Bayes-ToM$_1$ improved its utility as well as other metrics significantly compared to the agent equipped with the zero-order belief. This experiment verified that Bayes-ToM$_1$ was effective for advanced agents and it also took advantage of Bayes-ToM with a lower-order one.

5.3 New Strategy Detection

In this section, we evaluate the detection ability of Bayes-ToM$_1$. The opponent started with a policy that the Bayes-ToM$_1$ agent had already learned already, and after 5 episodes it switched to another policy that was not seen by the Bayes-ToM$_1$ agent (i.e., not included in the policy library). Besides, we also changed the switch episodes to 10 and 20 episodes to study how the switch period affects the detection performance (Fig. 3).

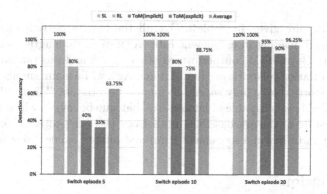

Fig. 3. The performance of our agent against new strategy

As the figure depicts, the Bayes-ToM$_1$ agent had a high-quality performance in all cases, and the averaged performance improved as the switch frequency decreased (e.g., when the switch period is 20, Bayes-ToM$_1$ can reach a accuracy of 100% for all opponents). The experiments showed that the Bayes-ToM$_1$ agent was able to detect opponent policy accurately.

5.4 Human Evaluation

To evaluate the practicality of our Bayes-ToM dialogue agent, we recruited 50 participants in the negotiation task. Human participants were first given a detailed tutorial of the negotiation. They then conducted the negotiations on computers via a negotiation websever. Four negotiation agents are used to match human players in this experiment, including Bayes-ToM$_1$, RL agent, ToM(e). Each human negotiator was partnered with one of the agents or another volunteer at random in order to compare the agents' performance to that of humans under the identical conditions. After a negotiation, human participants were asked to fill out a questionnaire, in which they answered questions regarding the following four aspects (using a 5-point Likert scale), i.e., intelligence of opponents (In), language fluency (Fl), language logic (Lg), willingness to play again (Wl). Averaging the four indexes, a comprehensive human-likeness score (Hu) can be obtained.

Table 3. The results of human-machine negotiation. The best value of each column is shown in **bold**.

	Ag	Ut	Fa	Len	Hu	In	Fl	Lg	Wl
Human	**0.76**	**0.56**	0.66	**12.2**	**4.93**	4.9	5.0	4.9	4.9
Bayes-ToM$_1$	0.68	0.5	**0.67**	12.9	4.4	4.7	3.9	4.3	4.7
ToM(e) agent	0.58	0.45	0.47	13.8	4.0	4.4	3.7	4.1	4.0
RL agent	0.61	0.33	0.29	14.3	3.5	3.9	3.2	3.5	3.4

The results are given in Table 3. Overall the human participants were the most successful, leading in agreement rate (Ag), dialogue length (Len), human likeness (Hu). Bayes-ToM$_1$ outperformed other agents and was ranked second in terms of negotiation related score (e.g., Ut and Ag), while maintaining reasonable human-like scores, being only second to human negotiators. The results suggest that Bayes-ToM$_1$ was capable of human-like dialogue behavior, which was better than other state-of-the-art baselines both from the perspective of negotiation outcomes (e.g., utility and agreement rate) and human likeness scores.

6   Conclusion

This paper propose the Bayes-ToM negotiation agent to model opponent strategy and select the optimal policy accordingly for negotiation dialogues. The extensive experimental results show that Bayes-ToM significantly outperformed state-of-the-art baselines on the CRAIGSLISTBARGAIN dataset. In addition, Bayes-ToM can also achieve human-likeness when negotiating with human negotiators.

The exceptional results justify to invest further research efforts into this approach. Regarding our future research, we plan to leverage prior knowledge from past learned policies of relevant tasks to accelerate learning with transfer learning. Moreover, we will explore possibilities to enlarge the scope of our approach towards other settings like multi-lateral and concurrent negotiation.

Acknowledgments. This study was supported by the National Natural Science Foundation of China (Grant No. 61602391).

References

1. Chandrasekaran, A., Yadav, D., Chattopadhyay, P., Prabhu, V., Parikh, D.: It takes two to tango: towards theory of AI's mind. CoRR abs/1704.00717 (2017)
2. Chen, S., Su, R.: An autonomous agent for negotiation with multiple communication channels using parametrized deep q-network. Math. Biosci. Eng. 19(8), 7933–7951 (2022)
3. Chen, S., Sun, Q., Su, R.: An intelligent chatbot for negotiation dialogues. In: Proceedings of IEEE 20th International Conference on Ubiquitous Intelligence and Computing (UIC), pp. 68–73. IEEE (2022)

4. Chen, S., Weiss, G.: An approach to complex agent-based negotiations via effectively modeling unknown opponents. Expert Syst. Appl. **42**(5), 2287–2304 (2015). https://doi.org/10.1016/j.eswa.2014.10.048

5. Chen, S., Yang, Y., Su, R.: Deep reinforcement learning with emergent communication for coalitional negotiation games. Math. Biosci. Eng. **19**(5), 4592–4609 (2022)

6. Chen, S., Yang, Y., Zhou, H., Sun, Q., Su, R.: DNN-PNN: a parallel deep neural network model to improve anticancer drug sensitivity. Methods **209**, 1–9 (2023). https://doi.org/10.1016/j.ymeth.2022.11.002

7. Crandall, J.W.: Just add pepper: extending learning algorithms for repeated matrix games to repeated Markov games. In: van der Hoek, W., Padgham, L., Conitzer, V., Winikoff, M. (eds.) International Conference on Autonomous Agents and Multiagent Systems, AAMAS 2012, Valencia, Spain, June 4–8, 2012, vol. 3, pp. 399–406. IFAAMAS (2012)

8. Haarnoja, T., Zhou, A., Abbeel, P., Levine, S.: Soft actor-critic: off-policy maximum entropy deep reinforcement learning with a stochastic actor. In: Dy, J.G., Krause, A. (eds.) Proceedings of the 35th International Conference on Machine Learning, ICML 2018, Stockholmsmässan, Stockholm, Sweden, July 10–15, 2018. Proceedings of Machine Learning Research, vol. 80, pp. 1856–1865. PMLR (2018)

9. He, H., Chen, D., Balakrishnan, A., Liang, P.: Decoupling strategy and generation in negotiation dialogues. In: EMNLP. Association for Computational Linguistics, pp. 2333–2343 (2018)

10. Hernandez-Leal, P., Kaisers, M.: Towards a fast detection of opponents in repeated stochastic games. In: Sukthankar, G., Rodriguez-Aguilar, J.A. (eds.) AAMAS 2017. LNCS (LNAI), vol. 10642, pp. 239–257. Springer, Cham (2017). https://doi.org/10.1007/978-3-319-71682-4_15

11. Hernandez-Leal, P., Taylor, M.E., Rosman, B., Sucar, L.E., de Cote, E.M.: Identifying and tracking switching, non-stationary opponents: a bayesian approach. In: Albrecht, S.V., Genter, K., Liemhetcharat, S. (eds.) Multiagent Interaction without Prior Coordination, Papers from the 2016 AAAI Workshop, Phoenix, Arizona, USA, 13, February 2016. AAAI Technical Report, vol. WS-16-11. AAAI Press (2016)

12. Keizer, S., et al.: Evaluating persuasion strategies and deep reinforcement learning methods for negotiation dialogue agents. In: Lapata, M., Blunsom, P., Koller, A. (eds.) Proceedings of the 15th Conference of the European Chapter of the Association for Computational Linguistics, EACL 2017, Valencia, Spain, April 3–7, 2017, vol. 2: Short Papers, pp. 480–484. Association for Computational Linguistics (2017)

13. Lee, S., Heo, Y., Zhang, B.: Answerer in questioner's mind: information theoretic approach to goal-oriented visual dialog. In: Bengio, S., Wallach, H.M., Larochelle, H., Grauman, K., Cesa-Bianchi, N., Garnett, R. (eds.) Advances in Neural Information Processing Systems 31: Annual Conference on Neural Information Processing Systems 2018, NeurIPS 2018, December 3–8, 2018, Montréal, Canada, pp. 2584–2594 (2018)

14. Lewis, M., Yarats, D., Dauphin, Y.N., Parikh, D., Batra, D.: Deal or no deal? end-to-end learning of negotiation dialogues. In: EMNLP 2017, pp. 2443–2453. Association for Computational Linguistics (2017)

15. Mnih, V., et al.: Asynchronous methods for deep reinforcement learning. In: Balcan, M., Weinberger, K.Q. (eds.) Proceedings of the 33nd International Conference on Machine Learning, ICML 2016, New York City, NY, USA, June 19–24, 2016. JMLR Workshop and Conference Proceedings, vol. 48, pp. 1928–1937. JMLR.org (2016)

16. Mnih, V., et al.: Human-level control through deep reinforcement learning. Nature **518**(7540), 529–533 (2015)
17. Roman, H.R., Bisk, Y., Thomason, J., Celikyilmaz, A., Gao, J.: RMM: a recursive mental model for dialog navigation. In: Cohn, T., He, Y., Liu, Y. (eds.) Findings of the Association for Computational Linguistics: EMNLP 2020, Online Event, 16–20 November 2020. Findings of ACL, vol. EMNLP 2020, pp. 1732–1745. Association for Computational Linguistics (2020)
18. Rosman, B., Hawasly, M., Ramamoorthy, S.: Bayesian policy reuse. Mach. Learn. **104**(1), 99–127 (2016). https://doi.org/10.1007/s10994-016-5547-y
19. Su, R., Yang, H., Wei, L., Chen, S., Zou, Q.: A multi-label learning model for predicting drug-induced pathology in multi-organ based on toxicogenomics data. PLoS Comput. Biol. **18**(9), e1010402 (2022). https://doi.org/10.1371/journal.pcbi.1010402
20. de Weerd, H., Verbrugge, R., Verheij, B.: How much does it help to know what she knows you know? an agent-based simulation study. Artif. Intell. **199–200**, 67–92 (2013)
21. Williams, J.D., Kamal, E., Ashour, M., Amr, H., Miller, J., Zweig, G.: Fast and easy language understanding for dialog systems with Microsoft language understanding intelligent service (LUIS). In: Proceedings of the SIGDIAL 2015 Conference, The 16th Annual Meeting of the Special Interest Group on Discourse and Dialogue, 2–4 September 2015, Prague, Czech Republic, pp. 159–161. The Association for Computer Linguistics (2015)
22. Wu, L., Chen, S., Gao, X., Zheng, Y., Hao, J.: Detecting and learning against unknown opponents for automated negotiations. In: Pham, D.N., Theeramunkong, T., Governatori, G., Liu, F. (eds.) PRICAI 2021. LNCS (LNAI), vol. 13033, pp. 17–31. Springer, Cham (2021). https://doi.org/10.1007/978-3-030-89370-5_2
23. Yang, R., Chen, J., Narasimhan, K.: Improving dialog systems for negotiation with personality modeling. In: ACL/IJCNLP, pp. 681–693. Association for Computational Linguistics (2021)
24. Yang, T., Hao, J., Meng, Z., Zhang, C., Zheng, Y., Zheng, Z.: Towards efficient detection and optimal response against sophisticated opponents. In: Kraus, S. (ed.) Proceedings of the Twenty-Eighth International Joint Conference on Artificial Intelligence, IJCAI 2019, Macao, China, August 10–16, 2019, pp. 623–629. ijcai.org (2019)
25. Young, S.J., Gasic, M., Thomson, B., Williams, J.D.: POMDP-based statistical spoken dialog systems: a review. Proc. IEEE **101**(5), 1160–1179 (2013)
26. Zheng, Y., Meng, Z., Hao, J., Zhang, Z., Yang, T., Fan, C.: A deep Bayesian policy reuse approach against non-stationary agents. In: Bengio, S., Wallach, H.M., Larochelle, H., Grauman, K., Cesa-Bianchi, N., Garnett, R. (eds.) Advances in Neural Information Processing Systems 31: Annual Conference on Neural Information Processing Systems 2018, NeurIPS 2018, December 3–8, 2018, Montréal, Canada, pp. 962–972 (2018)

Author Index

© The Editor(s) (if applicable) and The Author(s), under exclusive license
to Springer Nature Switzerland AG 2023
M. Yokoo et al. (Eds.): DAI 2022, LNAI 13824, p. 103, 2023.
https://doi.org/10.1007/978-3-031-25549-6

Printed in the United States
by Baker & Taylor Publisher Ser

Printed in the United States
by Baker & Taylor Publisher Services